Patrick Moore's Practical Astronomy Series

Other Titles in this Series

Digital Astrophotography: The State of the Art

David Ratledge (Ed.)

With 125 Figures

Springer

Cover illustrations: All images from the author. Background figure 3.3, Sun. Insets: top left: figure 3.9, Jupiter; top right: figure 3.14, Albireo; bottom left: 5.12, Andromeda Galaxy; bottom right: 5.13, Orion Nebula.

British Library Cataloguing in Publication Data
Digital astrophotography : the state of the art. - (Patrick
   Moore's practical astronomy series)
   1. Imaging systems in astronomy 2. Photography-Digital
   techniques
   I. Ratledge, David, 1945–
   522.6′3
   ISBN 1852337346

Library of Congress Cataloging-in-Publication Data
Digital astrophotography : the state of the art / David Ratledge (ed.).
      p. cm. — (Patrick Moore's practical astronomy series, ISSN 1617-7185)
   Includes bibliographical references and index.
   ISBN 1-85233-734-6 (alk. paper)
      1. Astronomical photography—Amateurs' manuals. 2. Photography—Digital
   techniques—Amateurs' manuals. I. Ratledge, David, 1945– II. Series.

QB121.D54 2005
522′.63—dc22   2005042544

Patrick Moore's Practical Astronomy Series ISSN 1617-7185
ISBN-10: 1-85233-734-6
ISBN-13: 978-1-85233-734-6
Springer Science+Business Media
springeronline.com

Typeset by EXPO Holdings, Malaysia
Printed in Singapore
58/3830-543210   Printed on acid-free paper   SPIN 10866042

Preface

In the years since *The Art and Science of CCD Astronomy* was first published, digital imaging has been transformed from what was, in reality, a minority interest to mainstream. Not even the most committed of CCD devotees could have predicted the few years it would take for digital imaging to supplant film. We all probably guessed that a new age was dawning, but the speed at which silicon sensors came to dominate the photography market was simply staggering. New areas also appeared. No one predicted webcams would become the instrument of choice for imaging the planets. Afocal photography re-emerged in digital format. For mainstream imaging, color has become almost the norm. It was therefore time for a new book – and one in color!

If you read the astronomical magazines, you are, no doubt, familiar with the names and images of our contributors. *Sky & Telescope, Astronomy, Night Sky, Astronomy Now* and other leading magazines from around the world have all included their work, in terms of both images for their gallery sections and feature articles.

The contributors have been selected for their expertise in a particular field although, in fact, most are multi-talented. First and foremost they are image takers – they are not writing about other people's images; they are writing about their own. You are hearing it from the horse's mouth! The big advantage of a book like this is that we have experts in each field rather than a single author who would perhaps be more familiar with some subjects than others. One person could never have the breadth of knowledge that we have incorporated here.

The book is divided into three sections, which broadly increase in sophistication and, unfortunately, in cost. The intention is to have something for every level of interest – and pocketbook! Topics range from using a consumer camera at the eyepiece of an ordinary telescope up to specialist multiple robotic telescopes searching for supernovae. Remember, even those with the most comprehensive setups started more modestly and got where they are today as their interest and knowledge developed over many years.

David Ratledge
Lancashire, UK

Contents

CHAPTER ONE

Introduction

David Ratledge

Background

There has never been a more exciting time to be an amateur astronomer. A new digital age has dawned, providing us with an arsenal of affordable imaging equipment, the power of which would have been unimaginable just 10 to 15 years ago. Those with the expertise and knowledge to exploit the digital tools are already reaping the rewards. This new generation is pushing forward further and further the bounds of what amateur astronomers can achieve.

At the heart of the digital revolution has been the silicon imaging chip. First on the scene was the CCD (charge coupled device) type but this has been joined, especially in the consumer market, by the CMOS (complementary metal oxide semiconductor) variety. Initially CCDs were only available in relatively expensive purpose-built astronomical cameras, but that has changed and even a humble $100 webcam can boast a supersensitive CCD chip at its heart. Consumer digital cameras are taking over amateur photography, making off-the-shelf multi-megapixel devices commonplace. Digital SLR cameras with interchangeable lenses are now available from virtually all big camera manufacturers. Their prices have dropped dramatically, and several are now around the $1000 price barrier. Even purpose-built cooled astronomical cameras have undergone a revolution, and the range now extends from $500 to more than ten times that. There is certainly a state-of-the-art imaging device available to suit all budgets. The technological advance sees no sign of coming to an end. Already the new kid on the block, the CMOS sensor, is widespread in the consumer market and is being used for astronomical imaging too. For those willing and able to embrace the new technology, then, the sky is literally the limit!

The Digital Era

Like many good inventions the CCD was actually the result of serendipity. Back in 1969 two research scientists at Bell Labs, Willard Boyle and George Smith, were trying to find a way to store data for computers – not an imaging device at all. The story goes that in just 1 hour of bouncing ideas off one another they had invented the CCD! But it gets even better – the storage device they came up with turned out to be sensitive to light! Within 5 years the first imaging chip had been produced (100 × 100 pixels), and the following year (1975) a TV camera had been equipped with one. Within 10 years of their invention one had been put to astronomical use. This early camera demonstrated the CCD's superiority over film, and the death of the photographic emulsion, at least for professional astronomy, would soon follow. However, it took until the 1990s for this revolution to reach mainstream amateur astronomers, with the last 5 years seeing an explosion in the number and types of cameras now available on the amateur market.

Initially, amateur CCD cameras appealed only to those comfortable with the emerging technology. However, the parallel arrival of everyday consumer cameras based on the same underlying technology has meant that digital imaging has become almost the norm. The user of a digital camera for holiday snaps will not be daunted by using similar technology for astro-imaging; in fact this new generation would *expect* to use a digital camera on a telescope. All indications are that the digital revolution in astronomy has only just begun.

The Technology

I have not yet explained what a CCD is. The device invented by Boyle and Smith was named a charged coupled device, hence CCD. It is, of course, a silicon chip – but one with a difference. Whereas most chips are covered in a black plastic casing to keep light out, a CCD chip has a window opening on its top especially to

Figure 1.1. CCD chip – note the window on the top surface to admit the light.

let the light in (see Figure 1.1). Silicon is sensitive to the visible and near infrared parts of the spectrum – outside this range light is either reflected off or passes straight through. What is meant by *sensitive* is that it will convert incident light (photons) into an electric charge (electrons).

The active light-exposed part of the CCD is divided into photosites or pixels in a matrix of rows and columns – a bit like a chess board with a multitude of tiny squares. I will use the term *photosite* in this introduction, but the term *pixel* is equivalent. Each photosite converts light (photons) into electrons and crucially stores them until the end of the exposure. The number of electrons produced is proportional to the light intensity. All we have to do is read out the electrons

Figure 1.2. State-of-the-art astronomical CCD cameras. Clockwise from top left: Starlight Xpress SXV-M25 – 6-megapixel APS sized single-shot color imager; Finger Lakes Instrumentation's high quantum efficiency (85%) entry-level ME2 camera; Santa Barbara Instrument Group's STL-11000M, 11-megapixel full frame imager; Apogee Instruments Inc.'s Alta E Series Internet remote-controllable camera.

from each photosite (square on our chess board), and we have the makings of a digital image. However, it is a bit more complicated than that! In a CCD, the value of a particular *x-y* photosite cannot be read out directly but can only be read out from the edge row. Each row is "coupled" (hence the name) to adjacent rows, and when one row has been read out, all the remaining rows are shunted down one, the next is read out, and so on.

A crucial factor in the superiority of CCDs over film is their quantum efficiency (QE). This is a measure of their efficiency in turning incoming photons into an electronic signal and is invariably represented as a percentage. A QE of 100% would be one where every incoming photon is detected and its effect is present in the output. The QE for film is of the order of only 2 to 3%, and even exotic treatments, such as hypersensitization, barely get it up to two figures. For daylight scenes that is not a problem but for astronomy and its low light levels that is profligate waste. The QE for CCDs, on the other hand, varies with wavelength but typically peaks at between 40 and 85%, and even at the blue end of the spectrum, where its efficiency drops off, it still comfortably exceeds that of film (see Figure 1.3). QE is often quoted by CCD manufacturers, but be aware that some quote relative QE and some absolute. *Relative* means the values given are a

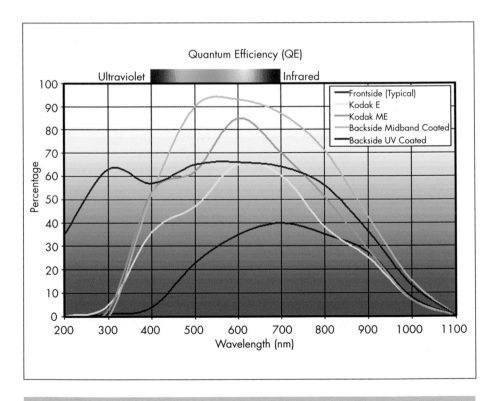

Figure 1.3. Chart showing typical quantum efficiencies for a range of CCD types. Note how the Kodak ME series almost matches the efficiency of the (expensive) back-illuminated CCDs.

proportion of the maximum efficiency of the device – so figures like 70% in the blue might mean 70% of the peak efficiency in say the red, which is actually a considerably lower QE. Bear in mind also that color CCDs have built-in filters that further reduce the QE. The underlying silicon might have a QE of 50% but the actual photons reaching the photosite will be substantially reduced before they can be detected. Inherent QE varies with the type of CCD, so it is appropriate to consider what those types are.

For amateur CCD cameras there have been three main types of CCD available. The first is the interline device. These are prevalent in domestic video cameras and are optimized for fast readout – essential for video operations. This is achieved by each column of active photosites (pixels) being paralleled by another shielded column of inactive photosites right next to it. Fast readout is possible because at the end of each exposure the charge (electrons) from each photosite is shifted at high speed into the adjacent shielded ones. These can be read out while the next exposure is taking place. Because only half of the columns of the device are being used for photon detection, their QE is immediately halved. Manufacturers can mitigate this by having micro-lenses focusing the light onto the active part but inevitably QE is going to be reduced. Nevertheless, for amateur astronomical CCD cameras interline devices are extremely useful and their mass manufacture for video and consumer digital cameras means they are an excellent value for money. They are the enabling technology behind the one-shot color camera, which makes color planetary imaging so simple and straightforward.

The most common type of CCD on the amateur market is still probably the front-side illuminated CCD. As might be guessed from their name, the light arrives on the front side of the chip. However, the front side is where the surface channels and gate structures are laid down. Beneath these lies the bulk silicon that will absorb the photons and generate the electrons. Therein lies the problem. The light has to pass the front structures *before* it can be recorded. That reduces the QE of the device as some photons don't make it through, but recent advances, such as the Kodak E and ME series, have reduced the losses. The original front-sided CCDs achieved QE of around 40%, the E series pushed this to more than 70% and the ME series (the M stands for micro-lenses) have lifted it up to 85%. QEs this high were previously the preserve of the back-side illuminated CCDs. Unlike interline CCDs, all the silicon in a front-side chip is available for receiving photons. However, in the ABG (anti-blooming gate) variety this is not the case. Blooming or bleeding is the ugly vertical streaks that occur from oversaturated bright stars and result from electrons spilling into adjacent pixels. ABG overcomes this by having vertical drains which, rather like the interline CCD, are inactive areas that allow the electrons to escape. This comes with a penalty, as not all the silicon is available for detecting photons. QE is reduced by around 25%. This is a heavy price to pay and there are alternatives, such as taking short exposures and summing them, that can reduce bleeding for non-ABG CCDs when bright stars are present.

Traditionally only affordable to professionals, back-sided illuminated CCDs by SITe and Marconi have entered the amateur market. Sometimes referred to as "thinned" devices, they are inverted after manufacture, exposing the bulk silicon directly. This process involves thinning them down to around 15 microns and re-mounting them upside down. Light no longer has to fight its way through the gate structures. As a result QEs have been the highest available and they are therefore highly desirable to both professional and amateur astronomers. The down side is that their manufacture is expensive. A feature of these CCDs is options on coatings. These can tweak the spectral response of the device. A midband coating produces an enhanced visible/infrared sensitivity with a QE peak greater than 90%. A broadband coating enhances the blue sensitivity, while one with a UV coating takes the spectral response into the ultraviolet.

CCDs are not the only type of solid-state imaging device. As already mentioned, the CMOS (complementary metal oxide semiconductor) imager has joined the party. It is still made of silicon so its intrinsic properties are similar to that of the CCD. They have the big advantage that their manufacture is not dissimilar to that of standard computer chips so they benefit from economies of scale as they are made in standard wafer foundries ("fabs"). Where they differ from CCDs is that on each photosite there are processing electronics such as transistors, amplifiers and circuitry – the actual amount varies according to type, but they are virtually a camera on a chip with the minimum of ancillary electronics needed. The readout procedure is simpler too, so much so that subsections and even single pixels can be read out – something not possible with CCDs. So is the CCD era coming to an end already?

Well, the answer is probably yes… and no. Certainly several areas where CCDs are in use today could well be replaced by CMOS technology. CMOS imagers will probably take over the low-cost high-volume market such as consumer digital still and video cameras. High-performance low-volume markets such as astronomical and medical imaging are likely, for the foreseeable future, to be the preserve of the CCD. Early CMOS imagers were bedeviled by noise and gained themselves a very poor reputation. Like all new technologies they have matured and improved. However, intrinsically they have a lower quantum efficiency (particularly in the red) because of that circuitry over the photosite. This can be mitigated to some extent by micro-lenses focusing the light onto the active part of each pixel. Whichever technology wins, the final product is the same, i.e., a digital image, and as far as we are concerned we could well have more choice and the prospect of falling costs.

Which Digital Camera to Buy

Which is the best digital camera to buy? A frequent question but one to which there is not one answer. It is equivalent to asking which telescope is best. Some telescopes are good for the planets, some are best for deep-sky. Others are optimized for portability, while some require an observatory. It is the same for digital cameras and the CCD chip at their heart. There is a further limitation in that the CCD should be optimized not only to the type of objects to be imaged but to the telescope as well. For many starting out in digital imaging who are asked what their target objects are, the answer is everything! That makes choosing a single camera a little difficult, and some compromises will have to be made.

The starting point for choosing a camera should always be the individual pixel (photosite) size. These are usually square, but not always for video-derived chips, and are measured in microns (1/1000 of a millimeter). Common sizes currently range from 6 to 24 microns (see Figure 1.4). As you might have guessed, generally

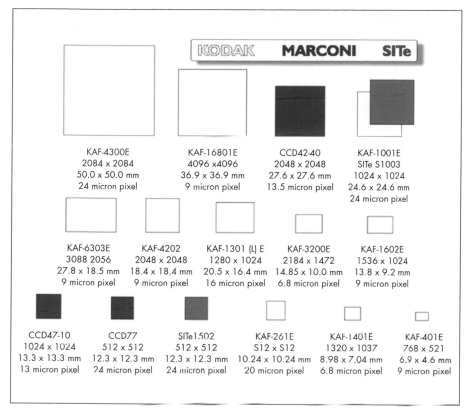

Figure 1.4. Sizes of common CCDs. (Courtesy Finger Lakes Instrumentation)

the low end ones are best for smaller telescopes and the high end for bigger telescopes. However, there is a bit more to it than that. It comes down to what is referred to as *sampling*, the actual number of pixels used to resolve detail. We need to be able to resolve the detail that our telescope and location are capable of delivering. Unfortunately, some math is needed to work this out. As a general rule we need to exceed the resolution of our telescope by a factor of at least 2, a bit more helps in the case of the planets.

So what is the resolution of your telescope and location? For long exposure deep-sky imaging from a suburban backyard, where atmospheric turbulence is the norm, it is probably only in the range of 3 to 5 arcseconds. Nothing like the theoretical figures quoted in telescope specifications! To exceed this by 2, we need a plate scale (a term from the old photographic days!) of say 2 arcseconds per pixel. To calculate this, use the formula:

plate scale (arcseconds per pixel) = 206.3 × pixel size (microns) / focal length (mm)

or

optimum pixel size = plate scale (arcseconds per pixel) × focal length (mm) / 206.3.

For a focal length of 2000mm and pixels of 20 microns, this computes to 2 arc-seconds per pixel – a good match for our telescope at its poor location. If a tele-compressor is used to reduce the focal length to around 1250mm then the result is 3.3 arcseconds per pixel, which is insufficient, and this would be referred to as undersampling. This would be okay for, say, supernova detection but the result-ing image would not be very photo-realistic – stars would be blocky and square. Undersampled images cannot be "improved" much by image processing so, gen-erally, it is better to err by over- rather than undersampling if "pretty pictures" are the goal. A better choice for the reduced focal length would be pixels around 9 to 13 microns.

The example given referred to deep-sky imaging. However, it is a somewhat different story for the Moon and planets. Here we will wait for optimum seeing, and exposures are so short that suburban seeing turbulence is much reduced. If we redo the math with the same 2000mm focal length, a typical telescope resolu-tion of around 0.5 arcsecond maximum gives a plate scale of 0.25 arcsecond per pixel. For this we find we need 2.4 micron pixels. Clearly we would need to magnify the image using eyepiece projection or a Barlow lens (i.e., increase the effective focal length) to match the typical pixel sizes available to us. A point to note from this is that for Jupiter, at 50 arcseconds maximum size, we will need a CCD with around 200 to 250 pixels across to resolve all the detail. However, a bit more helps, and oversampling is always better for the planets where we will want to "sharpen" the raw images considerably.

Color

We live in a colorful universe, so why not image it in color? Until recently, CCD imagers have only been monochromatic. What has changed this is the arrival of the X3 sensor from Foveon (strictly it is a CMOS imager not a CCD), which stacks three color-sensitive pixels one on top of another. For all other chips the only way to create a color image is by means of color filters. In a single-shot color imager each photosite has its own color filter built on top of it. So some will have blue, some green and the others red filters (some sensors use the com-plementary colors). These filters are arranged in a special repeating pattern across the chip and the camera software is able to change the discreet color values into a smooth color image (see Figure 1.5). For the more normal mono CCD we have to use 3 color filters, in turn, in front of the whole chip. The number of images we need to take is immediately tripled. The single-shot color imager might seem the answer to all our needs but the mono imager has the benefit that, when used in mono mode, it has no filters reducing the light reach-ing the chip or, when used with a filter, all pixels are recording an image. The single-shot color on the other hand, for mono work, is approximately 3 times slower or less efficient – even more when you consider that a typical unfiltered CCD is highly sensitive in the near infrared as well. So if your primary interest is deep-sky then a mono imager plus separate filters makes a lot of sense but if your interest is the planets or the brighter deep-sky objects then a single-shot color imager does make life easy.

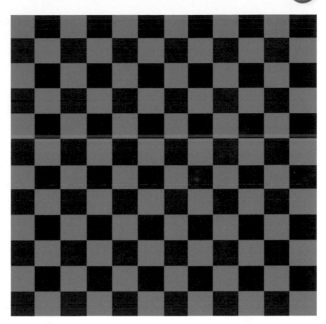

Figure 1.5. The color filter array (CFA) over the top of the sensor is needed so that it can create a color image. Without the CFA a sensor could only produce a grayscale image. In this array (Bayer) each sensor now only detects red, green or blue light.

For general deep-sky images a practical method of color imaging has emerged, known as LRGB (lightness or luminance, red, green, blue). Briefly this comprises adding, to our best monochrome image, the color information from a lower resolution one. This lower resolution image can be taken with the camera set to 2 × 2 binning, which will increase sensitivity but reduce exposures by 4 times. Alternatively, the color information can come from a different color CCD camera (if you have a friend with a Starlight Xpress) or even an old 35mm color slide.

Getting the Best Out of Our Camera

Probably the most important factor in getting the best out of our camera is that of maximizing the "signal-to-noise" ratio, S/N. A whole book could be devoted to this topic! It is frequently thought, particularly by those new to CCDs, that because they are more sensitive than film, only short exposures are needed. However, while it is true that an exposure of just a few seconds will produce some results and reveal the target object, there is a world of difference between such an image and one that has maximized the S/N. The former will be heavily speckled and gritty, the latter smooth and with a wealth of subtle detail. The reason is, of course, that all-important signal-to-noise ratio.

The signal part of S/N is the easiest to understand and is simply the number of photons recorded by the photosite or pixel. Noise is not quite as easy to grasp.

Here something called "uncertainty" rears its head. Detecting photons has an inevitable randomness – repeating the identical observation will not produce *exactly* the same numbers. This unpredictability, which can never be totally removed from a signal, is called *noise*. Note this subtle definition. An unwanted signal that can be removed is not noise. Dark current is therefore not noise – but the random element embedded in it is! This is a common misconception, which you will often see repeated. Now the good news: If we increase the signal by whatever means, such as a longer single exposure or multiple exposures, then the signal increases faster than the noise. We can fight back.

So what are the causes of noise in our image? We have already seen that just detecting photons has a randomness that creates noise. This noise is proportional to the square root of the signal. This means that as the signal increases the noise only increases as its square root. Therefore long exposures are best. It is hardly a new concept but if you want good images there is no short cut to long exposures. There are other sources of noise, which unfortunately further degrade our image, and we must be aware of and attempt to reduce them too.

The first is a part of the dark current count. Even when the telescope is capped, our CCD camera will produce electrons. This count is produced in proportion to the temperature of the CCD. The lower its temperature the lower the dark count. Remember the definition of noise – it is the unpredictable element that is the noise not the part that can be removed. For dark current the noise is again proportional to the square root of the count. This amount cannot be totally removed and it is therefore noise we must attempt to reduce. That is why astronomical cameras are cooled. Professional cameras are cooled to incredibly low temperatures in an attempt to get this as near zero as possible.

The second source of noise, important for us amateurs whose exposures tend to be short, is readout noise. Every time we read out an image from the CCD, that uncertainty rears its head again producing another source of noise. The good news is that modern cameras with their sophisticated readout software have reduced this considerably compared to several years ago. It is sensible to select a camera with a low readout noise so check those specifications. Because readout noise is not related to pixel size, larger pixels are actually better (because the signal will be higher), but of course we need to avoid undersampling. However, for amateur observers, readout noise is not usually the limiting factor unless very short exposures of faint objects are contemplated. The limiting factor is usually the next source – the curse of modern society!

Sky brightness is the most annoying source of noise. Just like dark current it produces an extra signal and one with an uncertainty. We can subtract an average value but we cannot subtract that uncertainty, which is again proportional to the square root of the signal. It gets worse. Sky brightness eats into the dynamic range of our camera. Each pixel has a maximum limit to its number of counts (full well capacity) and sky brightness added to the signal means that this limit will be reached much sooner. Again we get a commonsense result. Dark skies are really the best. Alternatively we can use filters that reduce sky brightness.

Finally there is image processing noise. It may come as a surprise but taking dark frames, bias frames and flat fields and applying them to our image all introduce noise. I am of course not advocating not taking them! Just be aware that we must take as much care over these frames as those of the actual image. A flat field from a month or two ago or a single dark frame rather than a median of many could actually make things worse. Take many calibration frames and average them to minimize that noise.

In summary, to maximize the *S/N* in our image:

- Maximize the signal. Longer exposures are an obvious solution but providing readout noise is not the limiting factor, so summing many exposures can be equally beneficial. A higher QE CCD such as a back-illuminated type is also best. Other helping factors are better tracking and focusing to concentrate those photons. A telescope with a bigger aperture or faster focal ratio helps. So does keeping the optics clean.

- Choose a camera with low readout noise. Readout noise is also minimized by longer exposures. If summing many exposures, these must not be so short that readout noise is the limiting factor. Again fast optics helps with this, as does bigger pixels.

- Select a camera with low dark current. Look for good cooling and perhaps the option of water cooling for even lower temperatures.

- Reduce the effect of sky brightness. Observe from a dark sky. If this is not possible use a filter. Look at shielding the telescope from direct light and minimize internal reflections. These don't stop sky brightness but if ignored would make the *S/N* even worse.

- Select a CCD that has pixels consistent with proper sampling. Remember small pixels can be binned to form larger pixels. This will improve the *S/N*.

- Take many dark, bias and flat field frames and average them to produce master frames. Regard 10 of each as an absolute minimum but the more the better so why not take 30 each!

Section 1

Getting Started

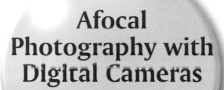

CHAPTER TWO

Afocal Photography with Digital Cameras

David Haworth

Introduction

Consumer digital cameras have revolutionized the ease of taking great pictures of the Moon and Sun (Figure 2.1). In no time at all you can aim your digital camera into the telescope eyepiece, take a picture, evaluate the result on the camera monitor and adjust the camera/telescope to improve your exposure. And of course you have the pleasure of seeing your results right away.

Digital cameras have no film and processing expense. As a result, the number of images is unlimited because you keep the best and delete the rest. This allows you to improve your imaging technique by experimenting with different settings and configurations. And most important, taking many images improves the odds of obtaining a stunningly detailed one during an instant of good seeing.

Digital camera images are easily enhanced with image processing software. By taking multiple images you can create lunar mosaics (Figure 2.2) or create time-lapse lunar and solar eclipse movies. Digital cameras provide a new exciting dimension to amateur astronomy.

Afocal Photography

The aiming and aligning of a camera lens into the telescope eyepiece is called afocal photography and can be used with any camera that has a lens (Figure 2.3).

The Moon is excellent for learning how to take afocal photographs. Start by using a long focal length eyepiece with a long eye relief and focus it on the Moon.

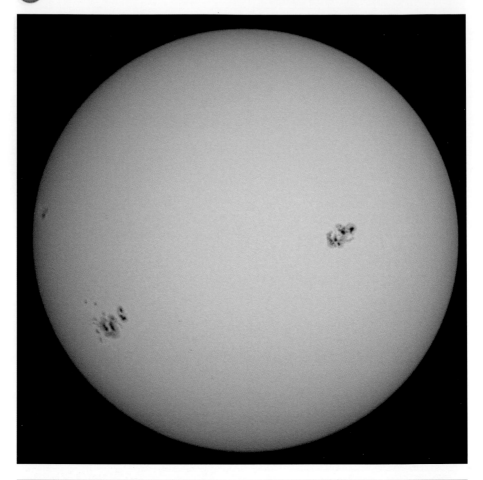

Figure 2.1. Sun image taken with a Nikon Coolpix 990 digital camera afocal coupled to a Tele Vue 55mm Plossl eyepiece, Tele Vue Paracorr coma corrector, Orion Telescope & Binoculars Atlas 254mm aperture reflector telescope, (f/4.7 focal ratio, 1200mm focal length) solar filter and Losmandy G-11 mount.

Next aim the camera lens into the telescope eyepiece. Center the camera lens over the eyepiece, being careful not to touch the camera lens with the eyepiece. Use the camera monitor to adjust both the position of the camera and the telescope focus. Press the shutter button while holding the camera steady. Take many images – remember they are free! For a typical night it is common to have one good image out of every 20 or more taken.

Next, take images with various camera settings, such as exposure time, zoom, manual operation and ISO values. Try various eyepieces to see which eyepiece combinations and camera settings produce the best images.

Figure 2.2. Moon mosaic from two afocal images taken with a Nikon Coolpix 990 digital camera coupled to a ScopeTronix STWA18 Wide Angle 18mm eyepiece/adapter, Orion Telescope & Binoculars Argonaut™ 150mm Maksutov-Cassegrain telescope (f/12 focal ratio, 1800mm focal length) and EQ-3 mount.

Afocal Photography Basics

Some math is required to get the best results, but don't worry; it is fairly easy to grasp. The formulas for afocal photography are listed in Table 2.1. The projection magnification is the camera lens focal length divided by the eyepiece focal length.

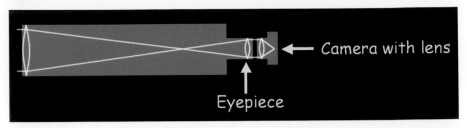

Figure 2.3. In afocal photography, the camera with its lens is aimed directly into the eyepiece.

Table 2.1. Afocal Photography Formulas

F_t	Telescope focal length
D	Telescope aperture diameter
f_t	Telescope focal ratio = F_t/D
F_{ep}	Eyepiece focal length
	Telescope exit pupil = F_{ep}/f_t
F_c	Camera focal length
f_c	Camera focal ratio
	Camera entrance pupil = F_c/f_c
M_p	Projection magnification = F_c/F_{ep}
$F_{effective}$	Effective focal length = F_t*M_p
$f_{effective}$	Effective focal ratio = f_t*M_p

The projection magnification can act like a Barlow when it is greater than one (shaded cells in Table 2.2) or as a focal reducer when it is less than one (clear cells in Table 2.2). Your telescope effective focal length is the telescope focal length multiplied by the projection magnification (Table 2.3). The telescope focal length and focal ratio are increased in the shaded Table 2.3 cells and decreased in the clear cells.

Table 2.2. Projection Magnification Factor

| Eyepiece Focal Length | Nikon 990 Camera Focal Length | |
	Minimum Zoom 8.2mm	Maximum Zoom 23.4mm
4mm	2.05	5.85
10mm	0.82	2.34
14mm	0.59	1.67
18mm	0.46	1.30
20mm	0.41	1.17
24mm	0.34	0.98
35mm	0.23	0.67
40mm	0.21	0.59
55mm	0.15	0.43

Table 2.3. Effective Focal Length and Ratio for a 1800mm Focal Length, f/12 Telescope

| Eyepiece Focal Length | Nikon 990 Camera Focal Length | | | |
| | Minimum Zoom 8.2mm | | Maximum Zoom 23.4mm | |
	Effective Focal Length	Effective Focal Ratio	Effective Focal Length	Effective Focal Ratio
4mm	**3690mm**	**24.6**	**10530mm**	**70.2**
10mm	1476mm	9.8	**4212mm**	**28.1**
14mm	1054mm	7.0	**3009mm**	**20.1**
18mm	820mm	5.5	**2340mm**	**15.6**
20mm	738mm	4.9	**2106mm**	**14.0**
24mm	615mm	4.1	1755mm	11.7
35mm	422mm	2.8	1203mm	8.0
40mm	369mm	2.5	1053mm	7.0
55mm	268mm	1.8	766mm	5.1

Things That Can Go Wrong

Camera movement and poor camera alignment over the eyepiece cause blurred images. As your exposure time increases, hand holding the camera steadily and accurately over the eyepiece becomes more difficult. To solve these problems the camera can be mounted on a tripod next to the telescope eyepiece, mounted on the telescope or mounted on the eyepiece. There are several options for mounting the camera: (1) use an adapter that mounts the camera on the telescope to position it over the eyepiece, (2) put an eyepiece in a universal camera adapter that is attached to the camera (Figure 2.4 and 2.5), (3) mount the camera onto an

Figure 2.4.
ScopeTronix 28mm camera to T-thread adapter, Orion Telescopes & Binoculars 20mm Sirius Plossl eyepiece and camera adapter before assembly.

Figure 2.5.
Assembled ScopeTronix 28mm camera to T-thread adapter, Orion Telescopes & Binoculars 20mm Sirius Plossl eyepiece and camera adapter.

Figure 2.6. Tele Vue 10mm Radian™ and 35mm Panoptic™ eyepieces with Tele Vue 28mm camera adapters.

Figure 2.7. The older Nikon Coolpix 990/995/(4500 not shown) digital cameras are excellent afocal photography cameras when used with eyepieces having 28mm camera threads. Eyepieces from left to right: ScopeTronix 14mm & 18mm, William Optics 24mm and ScopeTronix 40mm.

eyepiece using an adapter (Figure 2.6) or (4) mount the camera onto an eyepiece with a built-in adapter (Figures 2.7 and 2.8).

Another common problem is that some images may have dark corners or a dark circle frame (Figure 2.9). This darkening is called vignetting and occurs when the camera's imaging chip is not fully illuminated. Also, your images may have a dim fuzzy spot or circle in the center of the image. Both of these problems are the result of a poor telescope, eyepiece and camera optics configuration. You

Figure 2.8. The ScopeTronix 18mm eyepiece with 28mm camera threads attaches directly to the Nikon Coolpix 990 camera.

Figure 2.9. Daytime moon images showing vignetting at minimum zoom in the left image and removing it by increasing the camera zoom in the right image.

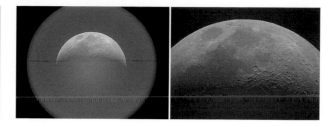

can reduce or eliminate these undesirable effects by using compatible telescope, eyepiece and camera configurations.

Your telescope and eyepiece have an exit pupil and eye relief. Your camera has an entrance pupil and a field stop. Typically, for best results, the telescope and eyepiece exit pupil should be equal to the camera entrance pupil, and the eye relief should be equal to the distance from the top eyepiece lens to the camera lens field stop. The camera lens field stop is where the camera iris diaphragm is located.

When the telescope and eyepiece exit pupil are larger than the camera entrance pupil, the camera field stop blocks some of the light and this light is not recorded as part of the image. The result is equivalent to using a smaller aperture telescope. Also, you might see a dim fuzzy spot or circle in the center of your image if you are using a telescope with a central obstruction like a Newtonian, Maksutov-Cassegrain or Schmidt-Cassegrain telescope.

The telescope exit pupil is the eyepiece focal length divided by the telescope focal ratio. For example, the exit pupil of an 18mm eyepiece used with a 150mm aperture Maksutov-Cassegrain telescope (f/12, 1800mm focal length) is 18mm/12 or 1.5mm.

The camera entrance pupil is the camera focal length divided by the camera focal ratio. A camera with a zoom lens has a range of entrance pupils. For example, the Nikon 990 zooms from 8.2mm to 23.4mm with focal ratios of f/2.5 to f/4, and its entrance pupil ranges from 3.3mm to 5.9mm (Table 2.4).

When using an 80mm aperture refractor telescope (f/5, 400mm focal length) with the 18mm eyepiece, some light is not imaged when the camera is set at its minimum 8.2mm zoom. At minimum zoom the camera entrance pupil is 3.3mm and will block some of the light because the telescope eyepiece exit pupil is 18mm/5 (3.6mm), which is 0.3mm larger than the camera entrance pupil. On the other hand, a 150mm f/12 Maksutov-Cassegrain telescope with the same 18mm eyepiece does not suffer from this problem because its 18mm/12 (1.5mm) exit pupil is smaller than the 3.3mm camera entrance pupil (Table 2.5).

Table 2.4. Nikon Coolpix 990 Entrance Pupils

Focal Length	Fastest Focal Ratio	Entrance Pupil
8.2mm	f/2.5	3.3mm
13.0mm	f/3.0	4.3mm
18.0mm	f/3.5	5.1mm
23.4mm	f/4.0	5.9mm

Table 2.5. Telescope and Eyepiece Exit Pupils Less Than the Nikon 990 Camera Entrance Pupil at Minimum Zoom

Eyepiece Focal Length	Telescope Focal Ratio		
	f/5	f/10	f/12
4mm	0.8mm	0.4mm	0.3mm
10mm	2.0mm	1.0mm	0.8mm
14mm	2.8mm	1.4mm	1.2mm
18mm	3.6mm	1.8mm	1.5mm
20mm	4.0mm	2.0mm	1.7mm
24mm	4.8mm	2.4mm	2.0mm
35mm	7.0mm	3.5mm	2.9mm
40mm	8.0mm	4.0mm	3.3mm
55mm	11.0mm	5.5mm	4.6mm

Note: Exit pupil = eyepiece focal length/telescope focal ratio. Shaded area shows a Nikon 990 camera is at minimum zoom and its 3.3mm entrance pupil is greater than the telescope eyepiece exit pupil.

When a 150mm Maksutov-Cassegrain telescope is used with a 55mm eyepiece, a large dim fuzzy circle appears in the image. It is caused by the telescope's central obstruction and the large 55mm/12 (4.58mm) exit pupil compared to the smaller 3.3mm camera entrance pupil at minimum zoom. In an extreme case, a camera entrance pupil could be so small that the camera field stop blocks all the telescope light and the camera entrance pupil sees only the central obstruction shadow. To fix this problem the camera zoom is increased until the camera entrance pupil is greater than the telescope's 4.58mm exit pupil. Vignetting changes with different zoom settings and it is usually reduced at maximum camera zoom.

Eye relief is the distance from the eyepiece lens to the point where your eye can best see clearly the full field of view. Your eyepiece eye relief should match the distance of the eyepiece outer lens to the camera lens field stop. Matching the eye relief to the camera field stop is difficult because the field stop distance is not specified by camera manufacturers, and in some cameras, like the Nikon Coolpix 990, it can change position with different zoom settings.

Cameras like the Nikon 990/995/4500 series with small lenses have a better chance of being optimally positioned with respect to the eyepiece eye relief when using an eyepiece with a long eye relief and a lens at the top of the eyepiece. See Afocal Photography Trouble-shooting (Table 2.6) for other problems, causes and solutions.

Digital Cameras

The Nikon Coolpix 990/995/4500 are great cameras for afocal photography. Their small lens provides a good optical match to eyepieces. They have automatic operation and full manual control of focus, aperture and shutter speed. Also, they

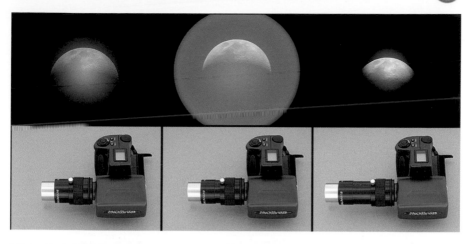

Figure 2.10. Adjusting the Nikon 990 camera field stop distance at minimum zoom to match the ScopeTronix 40mm eyepiece eye relief. From left to right: camera too close to the eyepiece lens, camera at best position and camera too far from the eyepiece lens. Images of the Moon were taken during daylight to show vignetting. With the ScopeTronix 40mm eyepiece you can adjust the distance of the camera from the top lens of the eyepiece.

have 28mm threads on the camera body so they can be mounted directly to eyepieces and eyepiece adapters. Other features include digital zoom for easier focusing, image histogram for accurate exposure adjustment, manual color balance, manual camera image processing, a remote shutter cable and a swivel camera body to position the camera monitor so that it is easily used when the camera is mounted on the telescope (see Figure 2.10).

I use the Coolpix 990 in manual exposure mode with the aperture set wide open and the shutter speed adjusted for correct exposure. Manual focus is set to infinity, and the flash is disabled. Camera sensitivity is best set to ISO 100 based on my experience with test images at ISO 100, 200 and 400. The higher ISO speed allows faster shutter speed but the images become very noisy. White balance is set to direct sunlight for all images. Contrast and brightness are set to normal, causing the same contrast and brightness adjustments on all images. Image sharpening is set to high and image size is set to the maximum of 2048×1536 pixels. Metering is not used.

Finally, the image quality is set to the highest-quality JPEG format. The camera can save images as TIFF, which gives better image quality than JPEG. However, the TIFF file size is approximately 9 times larger than the JPEG file so it takes much longer before the camera is ready to take the next image. I have found it is better to take many high-quality JPEG images and catch a moment of good seeing rather than take fewer TIFF images. Also, the quality difference between the JPEG and TIFF formats becomes less of an issue if you are reducing the images for use on a Web site.

Table 2.6. Afocal Photography Troubleshooting Chart

Problem	Possible Causes	Possible Solutions
Vignetting	Camera is not centered and perpendicular to the eyepiece optical axis.	Adjust the mounting of the camera above the eyepiece.
	Telescope exit pupil is larger than the camera entrance pupil.	Decrease the telescope exit pupil by decreasing eyepiece focal length or increase the telescope focal ratio.
		Increase the camera entrance pupil by changing the camera zoom.
		Move the camera closer or farther from the eyepiece so that the eye relief equals camera field stop distance.
		Use an eyepiece with longer eye relief.
		Use an eyepiece with a lens at the very top of the eyepiece.
		Use a camera with a small lens that has a shorter camera field stop distance.
		Focus the camera at infinity or use the camera macro mode.
Round dim spot in the center of images	Central obstruction is caused by the telescope exit pupil being larger than the camera entrance pupil.	Increase the camera entrance pupil by changing the camera zoom.
		Decrease the telescope exit pupil by decreasing eyepiece focal length or increase the telescope focal ratio.
	Central obstruction is caused by mismatch of eyepiece eye relief and camera field stop distance from the eyepiece.	Move the camera closer or farther from the eyepiece so that the eye relief equals camera field stop distance.
Blurry images	Camera movement or telescope vibration is caused by pushing the shutter button.	Mount camera on telescope or eyepiece, use remote camera shutter control or camera self timer.
	High magnification and a telescope that does not track the sky.	Decrease camera zoom, decrease exposure time or use a telescope mount that tracks the sky movement.

Table 2.6. Continued

Problem	Possible Causes	Possible Solutions
	Camera auto focus cannot focus.	Turn off auto focus, set the camera focus at infinity and focus with the telescope focuser.
		Image the Moon terminator that has bright and high contrast areas.
	Camera infinity focus setting is not correct.	Try other manual focus settings.
	Uneven focus across image.	Align the camera to be centered and perpendicular to the eyepiece optical axis.
	Poor focus.	Use maximum optical zoom and digital zoom in when focusing.
		Use focusing mask to find best focus.
		Use electric focuser to prevent telescope vibrations when adjusting the focus.
		Focus on high contrast object (Moon terminator, bright star, planet's moon, etc.).
		Use a magnifying glass to look at the camera monitor or large external monitor when focusing.
	Poor seeing (atmospheric distortions).	Take image when object is higher in the sky and do not take images over houses, hot pavement, etc.
		Decrease magnification by zooming out the camera or use an eyepiece with a longer focal length.
		Take many images to catch a good seeing moment.
		Decrease exposure time and increase camera ISO.
	Telescope, eyepiece or camera optics are hot.	Wait until optics cool down.
	Telescope vibrates because of the wind or breeze.	Use vibration suppression pads under the telescope mount and shelter the telescope from the wind.
	Small camera monitor is sharper than image.	Use external monitor with camera.

Table 2.6. Continued

Problem	Possible Causes	Possible Solutions
	Camera monitor shows a better quality image because it is a real-time stacked image.	Take an image and view single image with magnification on camera monitor.
Noisy images	Image is too dim.	Increase exposure time or stack images.
	ISO setting is too high.	Decrease ISO setting and increase exposure time.
	Camera noise for long exposure times.	Turn on camera noise reduction or take dark images and dark subtract the image.
	Camera is hot.	Turn off camera for several minutes between images to let it cool down and keep the camera monitor off as much as possible between images because it generates heat in the camera.
	Camera is sharpening the noise.	Decrease or turn off camera sharpening.
Dim images	Exposure time is too short or at night the image on the camera LCD monitor looks brighter than it is.	Increase exposure time and/or stack multiple images.
	Too long an effective focal length.	Reduce effective focal ratio.
Images with bright streaks coming from the very bright white areas.	Image is overexposed and blooming occurs.	Decrease exposure time or decrease ISO setting.
		Increase the effective focal ratio by increasing camera zoom.

Image Processing

Image-processing software enhances image details (Figures 2.11 and 2.12). How you view the image affects how you process it. For example, if you display your image on the Web, the image needs to be smaller than about 800 × 600 pixels to be Web site–friendly. This allows the complete image to be displayed on most computer monitors without scrolling. If you want to print the image, then process the image to have a resolution of 240 to 300 dpi and adjust the image to be brighter in the dark areas than a Web image because most printers will not print details in the very dark areas.

When sharpening the Moon, watch carefully for oversharpening and creating a bright edge at the limb. Cor Berrevoets' RegiStax (aberrator.astronomy.net/registax) wavelet image processing does a good job sharpening the image.

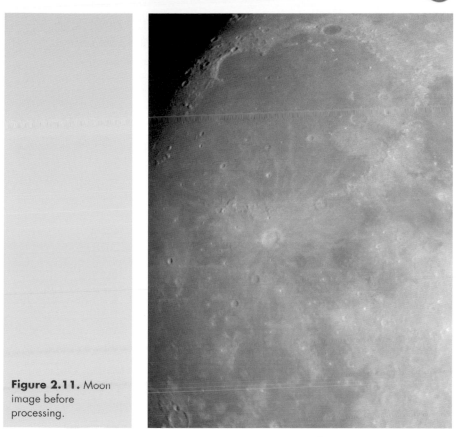

Figure 2.11. Moon image before processing.

The Nikon 990 full resolution is 2048 × 1536 pixels; its image needs to be resized 50% to 25% smaller to be Web site–friendly. Furthermore, as the image is resized smaller, the perceived image noise decreases and the image looks sharper.

Image Processing with Photoshop

With Adobe Photoshop you can sharpen the image's lightness channel after wavelet sharpening in RegiStax. Lightness channel sharpening does a good job of sharpening the image without causing color artifacts. There are three steps to this process. First, convert the image to Lab colors. Next, in Channels, select only the lightness channel and do an unsharp mask on it. Finally, convert the image back to RGB colors. The Moon's limb and image noise will determine the strength of lightness channel sharpening.

Levels are used to adjust the image's white and black points. Increase the contrast in the midrange with Curves. Be careful: with a too-strong "S" curve the bright and dark detail areas will be replaced by pure white and black.

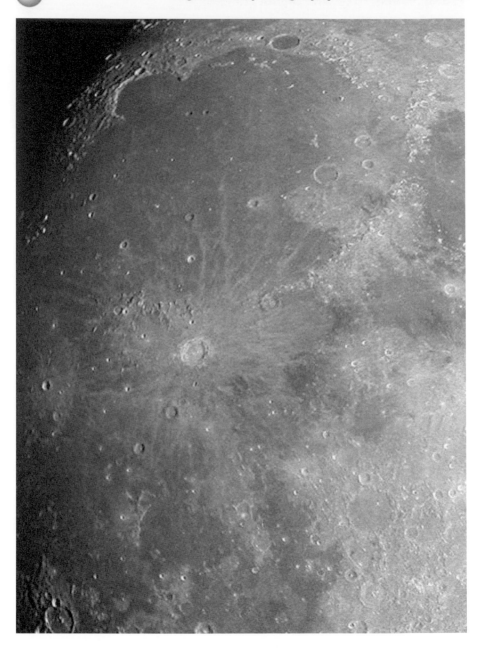

Figure 2.12. Moon image after processing.

Typically, you can improve an image after it is resized by sharpening it again. The brightening of the Moon's limb and dim noisy areas limit the sharpening for the whole image. Try sharpening only selective areas, not the whole image.

Figure 2.13. Sunset image was taken in Oregon, U.S.A., star party August 2003 with same equipment as in Figure 2.1.

When you have color fringing between the Moon's bright white limb and the black sky, select the black sky and expand this selection by a few pixels until just the black sky and the Moon's color fringing is selected. Decrease the color saturation for this selection in the Hue/Saturation menu until the color fringing goes away in the white areas of the limb.

If your image has color blotches, check each of the RGB channels to see if the color blotches are related to a noisy color channel. If so, select only the noisy color channel and Gaussian blur it. This technique was used to remove noisy blue color blotches in the sunset image (Figure 2.13).

Another technique is to oversharpen the image to the point that the dim areas become noticeably noisy, then select only the noisy dim areas and Gaussian blur them to reduce the noise.

Summary

Digital cameras make it easier for you take better images by providing immediate feedback. They allow you to take many exposures to catch that moment of good seeing, which is so essential for really great images. Afocal imaging gives everyone with a standard digital camera the opportunity to produce results far better than ever before.

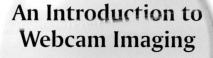

CHAPTER THREE

An Introduction to Webcam Imaging

David Ratledge

Introduction

"CCD cameras are too expensive for me" used to be a familiar story. All that changed with the advent of cheap video-conferencing cameras (Webcams). Models such as the Logitech (formerly Connectix), Quickcam, the Philips Vesta and Toucam and others from a variety of manufacturers have brought CCD imaging into the bargain basement. These $100 cameras take color images at typically 640 × 480 resolution and excel at the live display of telescope images to an audience. At last several people can "look" through a telescope simultaneously. But what really sets them apart is their ability to record tens of frames per second in the form of videos.

First things first – webcams are an excellent introduction to CCD imaging but they cannot compare with a $2000 SBIG! What they do better is record videos – and in color. Their refresh/display rate is fantastic so images appear continuously with very little delay. They are great for the Moon and planets, but their lack of cooling means that long exposures are compromised – only bright deep-sky objects such as the Orion Nebula are within range. But for the planets it is possible to collect hundreds of images in a matter of seconds. By selecting just the sharpest and adding them together the shortcomings of these cameras – high noise and low number of brightness levels – can be more than compensated for. They bring a new set of tools to the astro-imager.

The cameras are cheap because they do not have their own power supply and have no viewfinder or internal memory and therefore must be connected to a computer to function. For us this isn't a disadvantage but a big advantage. It means we can control and view images on a proper computer screen – not a tiny one on the back of a digital camera. In choosing a webcam, look for one with a CCD detector rather than the cheaper and generally noisier CMOS type. Also essential is control software that allows manual setting of the exposure. This can either be the software supplied in the box with the camera or special astronomical software such as IRIS (see later).

Quickcam

The Connectix Quickcam, probably the first webcam to be adapted for astronomical use, started the revolution. It is still available today (known as the Logitech Quickcam Pro) and has been upgraded to a USB interface. My introduction to the joys of webcams was with one of these first parallel interface cameras. The Quickcam proved the concept, but it had its faults. Chief among these was that it would "go to sleep" if the signal was not bright enough. This meant that unless you found the object quickly it would lock up. Shining a torch down the telescope would sometimes wake it up, but it was time for an upgrade. Enter the Philips Toucam.

Philips Toucam Pro (Models PCVC740K and PCV840K)

The Toucam promised (on the box) 1280×960 color resolution and fast USB download with no lockups! The resolution turned out to be interpolated – the CCD was actually a Sony HAD device with 640×480 pixels and a size of 3.6×2.7 mm. It is a common advertising ploy with webcam manufacturers to claim all manner of resolutions, but it is the resolution of the CCD chip that counts – not the output image resolution. However, it was a major step forward in both operation and image quality.

An essential feature of any webcam is manual control of the exposure settings. Surprisingly some early commercial astronomical video cameras did not have this. The Toucam has manual adjusters for exposure (longest 1/25th second), gain, brightness and saturation. It has an "auto" mode too, which is generally about right for the Sun and Moon.

As with any webcam the first hurdle to overcome is finding a way of connecting it to the focuser. The lens on the Toucam actually screws off so it was relatively simple to make a 1.25-inch adapter (see Figures 3.1 and 3.2). Such has been the explosion in webcam use for astronomy that several manufacturers are now supplying these adapters off the shelf.

Late in 2003, Philips released an upgraded version of the Toucam Pro II (Model PCVC840K), and there was much concern in the webcam fraternity that it would not be as good for astronomy. However, the camera is basically the same but housed in a smarter case. The sensor appears to be identical, i.e., a Sony 640×480 CCD and the software is still limited to a 1/25th-second longest exposure. On the downside the USB cable has been reduced to 1.5 meters long – half of what it used

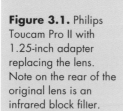

Figure 3.1. Philips Toucam Pro II with 1.25-inch adapter replacing the lens. Note on the rear of the original lens is an infrared block filter.

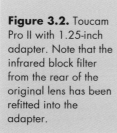

Figure 3.2. Toucam Pro II with 1.25-inch adapter. Note that the infrared block filter from the rear of the original lens has been refitted into the adapter.

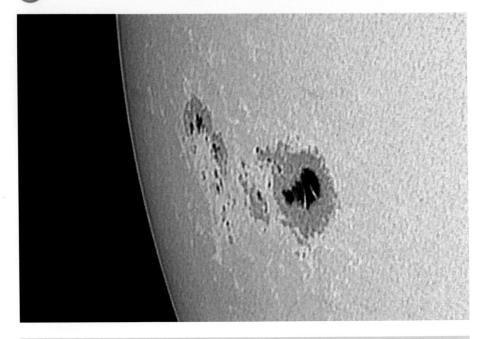

Figure 3.3. Solar closeup. Taken with a Toucam Pro II on a Celestron C8 at prime focus (2032mm focal length) and a Thousand Oaks solar filter. One of the problems with solar imaging is seeing the computer screen in broad daylight. A black cloth or hood helps.

to be. A USB extender cable will therefore generally be required (see Figure 3.4). The annoying red LED is also there, and the price went up slightly. Otherwise it was business as normal.

For centering and reaching focus it is best to have a wide-field eyepiece and have it approximately parfocal with the webcam. This can be achieved by taping the barrel of the eyepiece so it slides in to the precise common focus. Finding an object is fairly simple using an eyepiece and then switching over to the Toucam without disturbing telescope pointing. This way it will be close enough to focus to generate an image on the computer screen.

Having used an adapter and realized the power of the Toucam, I had more ambitious plans, and a second camera was purchased. My idea was to totally dismantle the camera and mount it on the back of an old SLR camera body. This would provide:

- Viewfinder for easy location and centering of bright objects.
- Visual setting of the approximate focus position.
- Camera lens mount for easy attachment of adapters or lenses.

The camera comprised a single circuit board with the CCD mounted fairly centrally. An old Pentax camera body was available for mounting the circuit board, which just needed the microphone disconnecting as it would have prevented it being mounted near to focus. It was simply spot epoxied into place and sealed

Figure 3.4. Toucam Pro II attached to the visual back of a Celestron C8 – the base of the webcam has been unclipped. Due to a shorter USB cable the laptop PC has to be close to the telescope. If you need to be further away, use a USB extender.

with silicone. After 24 hours it was ready for use (see Figures 3.5 and 3.6). I have subsequently mounted 3 more on the back of SLR cameras for friends. On these later conversions, to get the chip at focus so camera lenses can be used, I have ground recesses into the camera's film rails so the circuit board can sit lower. I

Figure 3.5. "Pentax" Toucam – front view with the camera mirror raised.

Figure 3.6. "Pentax" Toucam on 16-inch Newtonian. The CCD could be regarded as "air-cooled," which does improve performance!

aim to get the chip either at focus or fractionally inside focus. This way lenses will always reach infinity focus.

First testing with the Pentax Toucam was a revelation. A camera cable release was needed to raise and lower the flip mirror but everything had gone well – the chip was almost exactly central and near enough the right focus position. The viewfinder works brilliantly, enabling easy centering of objects on the CCD.

Imaging Procedure

For controlling the camera, I use Christian Buil's IRIS program for both video and stills. The software supplied with the camera works but it is not designed for astronomical imaging and is best left in the box. IRIS (see Chapter 6) is freeware and includes extensive image processing functions. All that is generally needed is to set the exposure manually, using slider bars, until all the detail is visible and

Figure 3.7. Composite image of Saturn and four of its moons taken by Keith Mallelieu and David Ratledge. Saturn imaged using a Celestron C11 and its moons with a 16-inch Newtonian. For Saturn, the best 150 frames from a 600-frame sequence were processed using Registax. Wavelet filters were used to sharpen the image and reveal traces of Enke's division.

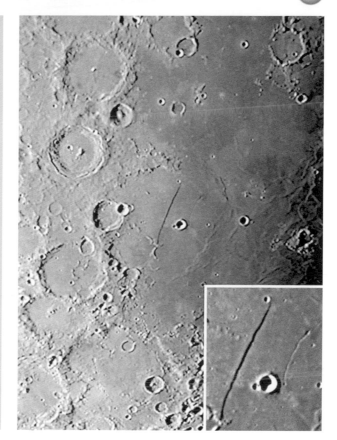

Figure 3.8. The Straight Wall on the Moon: main image best frames from 4 videos taken at the prime focus of a 16-inch Newtonian. Inset: best frames from 5 videos with 3× converter lens. Both sharpened with a combination of unsharp mask and wavelet filter in IRIS.

then perhaps tweak the brightness a touch. It is important to avoid overexposure – burnt-out detail is lost forever.

For the planets I connect a 3× converter lens on the camera body, and this provides an image scale of around 250 pixels for the size of Jupiter's disk. In the case of Mars, I use a 3× and 2× in tandem. For the Moon, either prime focus or a 2× converter is generally enough magnification (see Figure 3.8). Focusing on the planets can be difficult at high magnification because of seeing fluctuations but if the Moon is nearby its bright limb can be used to find precise focus.

Initially I took stills of the planets and the Moon. However, I soon learned that video is better – much better. With video, hundreds of images can be captured in seconds! The limit becomes disk space as 200 images at 640 × 480 equates to about 90Mb. Before you know it 1Gb of disk space can have vanished! Generally, though, the more images there are to select from, the better the result will be. For Jupiter the limit becomes how long before rotation of the planet starts to cause blurring (see Figure 3.9). I personally limit videos to no longer than one minute in which time, at 10 frames per second and allowing for some losses, 500 images can be collected. From these, the best 50 can be selected automatically during processing.

Figure 3.9. Jupiter and 2 moons – best frames from 3 videos. Processed using IRIS, which automatically selected the best frames.

Processing

For processing the videos I initially used Astrostack, a simple program to learn. It is freeware and does as its name suggests, stacks images together to improve image quality (signal to noise ratio). It takes video (avi) files and registers all the images in a sequence so even if your drive is not perfect, and the target has wandered around a bit, it will correct for this. It has limited image processing, but once stacked the image can be "improved" elsewhere by further processing. This method, however, assumes all images in a video sequence are good enough to be used. The reality is somewhat different and better results can be obtained by taking a longer video sequence and then editing out all the poor frames. The software IRIS can do this automatically ("Align&Stack") and it is now my preferred software for both image taking and image processing. There are several alternatives such as K3CCDTools and Registax. Both are freeware and provide automatic/manual selection of the best images from a video (avi) sequence. Registax is particularly informative as to what it is doing and permits various degrees of user intervention (see Figure 3.7). My suggestion is to try them all and choose the one that works best for you.

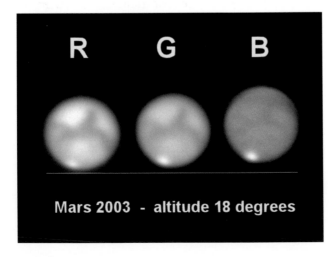

Mars 2003 - altitude 18 degrees

Figure 3.10. Mars – RGB planes shown to scale. The vertical displacement is caused by the 3 color planes being differently refracted through the atmosphere. Most software allows this to be corrected during processing.

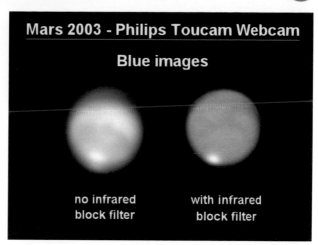

Figure 3.11. Blue images of Mars 2003 without and with infrared block filter.

IRIS does not operate on color images in one operation but splits them into red/green/blue files, which then have to be processed individually before recombining them to form a color image. In theory, this is best, especially for typical northern latitudes where the planets are never overhead. The reason for this is that the three color bands are refracted through the atmosphere slightly differently so they are not quite in perfect registration (see Figure 3.10). Final processing can then correctly realign the color planes. The 2003 image of Mars split into RGB planes demonstrates this clearly. Mars was only 18 degrees above the horizon and the 3 planes were markedly at different levels.

Another issue that the 2003 opposition of Mars demonstrated was the need for an infra-red block filter (see Figure 3.11). The Toucam lens incorporates such a filter so when we remove the lens we remove the filter too (see Figure 3.12). The effect of this is infra-red light leakage – the blue plane especially is affected by this and produces a double image. The improvement in sharpness and color fidelity when using an infra-red filter is dramatic.

IRIS has a single command, which selects the best frames from a sequence, aligns them and then adds (stacks) them together (see Figure 3.12). This makes processing very painless but the operation can take up to 30 minutes of computing

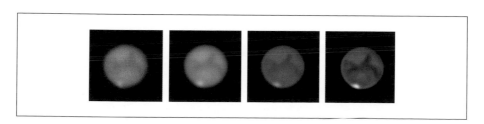

Figure 3.12. Mars 2003 – shot when only 18 degrees above the horizon. From left to right, single frame from a video; best 50 frames from a video of 250 frames; ditto sharpened; composite from 6 videos. Note the blue haze over the northern pole.

Figure 3.13. Crescent Moon Mosaic by Keith Mallelieu and David Ratledge, taken using a Philips Toucam with an 8-inch Newtonian (2× converter). The 27 sections were assembled using MAXIM. Full image (much reduced here) is 8 megapixels and has been printed more than 3 feet (90 cm) high.

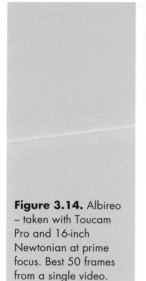

Figure 3.14. Albireo – taken with Toucam Pro and 16-inch Newtonian at prime focus. Best 50 frames from a single video.

time and the working files produced can require several gigabyles of spare disk space! Registax is somewhat similar but produces interesting graphs/charts as processing progresses.

For covering the Moon, mosaics of several images can be assembled (see Figure 3.13). This enables even the smallest chip to produce a megapixel image! Not only that – with each image using only the central on axis field the quality is the best your telescope can deliver. It is desirable to expose all the sections with identical exposure settings. This makes blending one to another much easier. For the Moon this means choosing an exposure that doesn't burn out on the brightest features and using that for all sections. To assemble the Mosaic, IRIS has a manual command (offsets have to be entered), but I prefer Maxim as it is so much easier, with images being overlaid by eye. I usually try for an overlap of about 100 pixels.

I also have tried deep-sky objects but, without modifying the electronics of the camera (see Chapter 4), this is limited to the brightest Messier objects – that 1/25th-second maximum exposure is just too short (see Figure 3.15). However, without the modification it is ideal for double stars. Traditionally, double stars have been difficult to photograph or image but using a webcam and shooting a video sequence the seeing can be (almost) frozen and excellent results obtained. The double star images shown in Figure 3.14 were obtained by shooting around 200 frame videos and selecting only the best. I found it advantageous to turn up the gain setting on the camera. IRIS was used to register and combine the images before slightly sharpening using its Richardson-Lucy routine.

Another use for webcams is guiding. For training the periodic error correction (PEC) of my drive I use IRIS and my webcam. This is so much easier than continuously looking through the eyepiece at an illuminated reticule. If the computer used for this is connected to the telescope and is able to issue commands to move

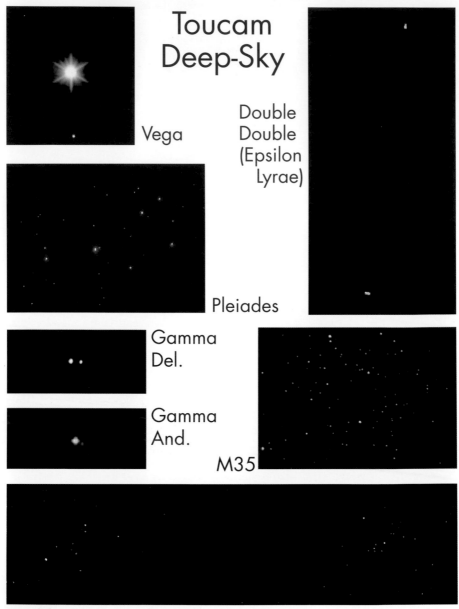

Figure 3.15. Double stars and deep-sky objects taken with an unmodified Toucam on a variety of telescopes and telephoto lenses.

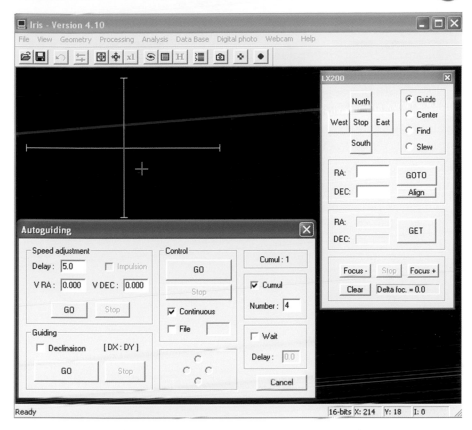

Figure 3.16. Autoguiding with a Webcam using IRIS. Note how the brightest star is shown as a cross rather than as an image. This makes manual guiding for PEC training very easy.

it, then auto guiding with the webcam is also possible. IRIS includes the commands to do this for LX200 compatible telescopes, and it can autoguide in RA and DEC or RA only (see Figure 3.16).

Conclusion

Webcams put the fun back into astronomy. Yes they are cheap and cheerful but they now probably produce the best-ever images of solar system objects. They also enable group viewing through a telescope, which is much preferable to queuing on open viewing nights. If you haven't tried them yet, why not give one a go. You will not be disappointed.

Useful Web Sites

Quickcam and Unconventional Imaging Astronomy Group (S.J. Wainwright)
http://www.qcuiag.co.uk/

Christian Buil's IRIS Software
http://www.astrosurf.com/buil/us/iris/iris.htm

Peter Katreniak's K3CCDTools Software
http://www.pk3.org/K3CCDTools/

Cor Berrevoets' Registax Software
http://aberrator.astronomy.net/registax/

R.J. Stekelenburg' s Astrostack Software
http://www.astrostack.com/

CHAPTER FOUR

Long-Exposure Webcams and Image Stacking Techniques

Keith Wiley and Steve Chambers

Introduction

In recent years a new method for astrophotography, which uses very afford-able equipment and produces competitive results, has emerged (see Figure 4.1). Standard webcams, originally created for the consumer-oriented video-conferencing and cheap home-based photography markets, have turned out to be an excellent source of affordable astrophotography cameras. These cameras often cost between $50 and $150 and, even after the modification described here, are still highly affordable compared to the astrophotography-oriented CCD cameras available from mainstream companies.

Planetary imaging can be performed with the right webcams without any modification, and you can start immediately if that is your interest. For deep-sky imaging, webcams out of the box suffer from several problems, most of which can be minimized or eliminated.

Astronomical Cameras and Webcams

The first mass-produced webcam, the black and white Quickcam made by Connectix, shared many similarities with dedicated astronomical CCD cameras in production at the same time, including the type of CCD used and the control method, which used the computer's printer connector to directly control the

Figure 4.1. M27, Dumbbell Nebula. Captured with an SC3 color Vesta and an SC3 B&W Vesta by Carsten Arnholm with a C8 scope and a Mogg .6× focal reducer. 65 × 40 sec color, 40 × 40 sec B&W. K3CCDTools, Registax2 and Photoshop.

sensor. Sadly the software that was supplied with the webcam had a bug that prevented the camera taking the long exposures required for deep-sky imaging. Dave Allmon started the first wave of interest in deep-sky webcam imaging by writing a control program for this web camera that allowed long exposures and the imaging of faint objects. Dave's first deep-sky images were of the Andromeda galaxy, with many fainter objects such as the Horse Head nebula soon following.

From this starting point, developments in webcams and astronomical CCD cameras have driven these products in very different directions. Astronomical CCD cameras have maintained a relatively high price and have aimed for precise digitization of the signal from each pixel of the CCD at the expense of image download time. Webcams have strived to achieve high frame rates of brightly lit subjects from color CCDs, quite often at the expense of image quality. To use modern webcams for deep-sky imaging we need to address a number of issues including exposure time and image quality. The solutions are a combination of hardware and software modifications and specialized image processing techniques.

Modified Cameras

The specific details of the modifications are best gained from the Web sites given in Table 4.1. Here, we will only cover the subject in broad terms. Also bear in

Table 4.1. Web sites with background information on long-exposure modified webcams.

http://www.qcuiag.co.uk/

http://groups.yahoo.com/group/QCUIAG/

http://www.firmament.tk/

http://www.pmdo.com/wintro.htm

mind that astronomical cameras based on these designs are available commercially if soldering to surface mount components is not one of your skills.

The first stage is acquiring a suitable webcam. While this sounds trivial, time spent researching the best webcams for modification will be very worthwhile. At the time of writing the Philips ToUCam II Pro is regarded as the best with the Logitech Quick Cam Pro 4000 and Creative's Ex Pro also worthy. All these cameras feature CCD sensors rather than CMOS devices. While some digital camera CMOS chips have been found to be very capable of astro-imaging, those currently used in webcams invariably have low sensitivity. This might change in the future, so it is worth checking Web sources such as QCUIAG to find the currently favored webcams.

Without modification, the cameras are limited to relatively bright objects such as the Moon and major planets. This lack of sensitivity is not due to inherent limitation in the CCDs used in the webcams. Indeed, the quantum efficiency of the best webcam CCDs is on a par with dedicated astro cameras. The sensitivity problem is linked to the webcam producing a moving image consisting of at least 5 frames per second. To image deep sky objects we really need to take pictures with exposures of 30 seconds or more rather than the 0.2 second offered by the standard webcam. The solution developed by the authors was to place some of the webcam's internal timing directly under the control of the PC. The electronic circuits needed to achieve this are very simple, but working on the surface mount components requires some skill and experience with a soldering iron. Good advice if you are considering doing this work yourself is to find a scrap circuit board with some surface mounted chips to practice on. Because the specifics of the modifications vary from camera to camera, these are best referenced from the Internet (see Table 4.2). Once modified, the webcam's exposures can be set to any length the user wishes by using software that is compatible with these modifications.

The modification to control length of exposure is fundamental to adapting webcams for deep-sky use. Further modifications to the camera's hardware are optional but offer additional benefits. The CCDs used in the webcams feature on-chip amplification circuitry. This significantly increases the quality of the images produced as it keeps the amount of electrical noise added to the image to

Table 4.2. Sites giving camera-specific modification details.

http://www.philip.davis.dsl.pipex.com/tcp2_mods.htm	ToUCam 2 by Phil Davis
http://mypage.bluewin.ch/bm98/l3k/modification.htm	QC3000 by Martin Burri
http://www.foley-tax.com/Astro/modz/	Creative Ex Pro by Jack Reed
http://www.pmdo.com/wwhich.htm	List of webcam modification sites

Table 4.3. Web sites detailing CCD replacement modifications.

http://www.astrosurf.com/astrobond/ebvpnbe.htm	Use of 1/4-inch B/W CCD. Etienne Bonduelle
http://www.pmdo.com/wsc3.htm	1/3-inch CCD. Steve Chambers
http://www.greg.beeke.btinternet.co.uk/icx414.htm	1/4-inch CCD. Greg Beeke

a minimum. However, this amplifier also emits photons by a process termed *electro-luminance*. In exposures lasting 30 seconds or more this results in an objectionable glow in a corner of the image corresponding to the part of the CCD array closest to this circuit. A solution is to drop the voltage to this circuit while the CCD is collecting its image and then to restore it when it is needed for reading the image out. This modification is, slightly confusingly, known as the "amp off" or "amp switch" modification. Also possible, although not particularly popular, is a modification to the webcams that allows half the CCD to be read out at a different speed from the rest of the array. This can be used to simultaneously guide a telescope using exposures of around one second while the main image is built up for a minute or two.

CCD-based webcams tend to use 1/4-inch color CCDs. The color information is gained from tiny red, green and blue filters that are located on top of the CCD structure. The arrangement is in blocks of 4 pixels, each having 2 green, 1 red and 1 blue filter. This allows the webcam to take a full color image without requiring external filters but, as the filters only let though a single color, the sensitivity is significantly reduced. Recently, amateur astronomers have successfully replaced the standard CCD with unfiltered black-and-white versions (see Table 4.3). It is also possible to swap the standard 1/4-inch CCD for one rather larger like a 1/3- or even a 1/2-inch CCD. When bought in small quantities, CCD chips can be quite expensive, so a 1/2-inch CCD will probably cost more than the original webcam. However, as the light gathered is proportional to the surface area of the sensor, these chip-swaps can be very worthwhile.

The final stage of modifying a webcam is often building a new case for it (see Figures 4.2, 4.3, and 4.4). While it is possible to fit a modified webcam back in

Figure 4.2. SC modded ToUcam, by Ashley Roeckelein.

Figure 4.3. Cooled SC modded ToUcam, by Ashley Roeckelein.

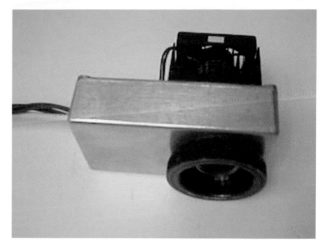

its original case, there are advantages to using a bigger box that allows for better air circulation and some cooling. When a webcam is left running, it consumes electrical power and produces some heat, which, if left to build up in the camera, will raise the CCD temperature and increase the thermal noise it produces. Simply allowing air to circulate and take away this heat is surprisingly effective, especially on a cold night. For greater cooling, Peltier coolers are able to reduce chip temperatures to 40 degrees or more below ambient. This does generate a whole new set of problems in stopping the chip from dewing or even icing up.

The standard lens in a webcam typically has a focal length of 7mm and a focal ratio of about f/3. This is ideal for capturing whole constellation pictures, maybe including some foreground subject. To move onto specific deep-sky objects a method for coupling the camera to a telescope is required. Probably the most versatile method is to incorporate a camera macro extension ring into the box design because this will allow both 1.25-inch focuser adaptors to be

Figure 4.4. Modded Vesta, by Keith Wiley.

Figure 4.5. M16, Eagle Nebula. Captured with an SC1.5 Vesta by Keith Wiley with a Meade 8″ f/6.3 LX200 and Mogg 0.6× focal reducer. 44 × 90 sec. Keith's Image Stacker and Photoshop 7.

used for prime focus telescope imaging and camera lenses to be employed for wide-angle shots.

After modifying the webcam hardware for use in deep-sky imaging, it is also possible to change the settings for the camera to make the best use of these alterations. The standard drivers for webcams tend to concentrate on producing the best high frame rate images of brightly lit subjects by using high compression and artificially emphasizing edges in images to give the appearance of sharpness. By writing directly to the memory chip within the webcam, it is possible to override these settings in favor of ones that are able to better represent the image formed on the CCD. This has been shown to give a very great improvement for webcams retaining their standard color CCD, with image artifacts much reduced. For webcams with black-and-white unfiltered CCD, the improvements are even better as higher resolution processing techniques can be employed (see Table 4.4).

Capturing Images

A webcam that has been modified to take long exposures doesn't work very well anymore with off-the-shelf webcam software. Astrophotography is much more

Figure 4.6. M51, Whirlpool Galaxy. Captured with SC3 B&W Vesta and SC color Vesta by Etienne Bonduelle with a Meade LX-90 scope and a 0.63× focal reducer. 53 × 90 sec IRB (Infared Black), 42 × 60sec IRB. Astrosnap, Registax, Iris and Paint Shop Pro 7.

Table 4.4. Web sites describing methods for altering the webcam's factory settings.

http://groups.yahoo.com/group/twirg/ http://groups.yahoo.com/group/qcuiag/	Yahoo discussion groups on webcam reprogramming
http://www.foley-tax.com/Astro/modz/Advanced.htm#ULTRA	Jack Reed's guide to webcam firmware
http://www.astrosurf.com/astrobond/ebrawe.htm	Etienne Bonduelle's guide to "Raw Mode"

demanding for a number of reasons. The first is that ordinary software no longer drives the camera properly because it has no way of controlling the new exposure control circuit.

There are a variety of freeware and shareware programs that control cameras modified according to our design. These programs go to great length to acquire the cleanest data possible by providing histograms that allow you to ensure that you are not saturating any part of the image and by performing no compression on the image before it is saved to disk. The most common capture programs are listed in Table 4.5.

The actual task of capturing images requires some understanding of how such images will be processed at a later time. Unlike film astrophotography or most professional CCD astrophotography, webcam astrophotography requires a process called image stacking, explained in greater detail later. For now, realize that image stacking means capturing numerous virtually identical images of an object during one session. Consequently, another function of the programs mentioned is batch image capture, in which a series of long exposures are captured one after the other, perhaps for an hour or more at a time.

Once your webcam is centered on the object in question, the next step is, of course, acquiring your focus. Focusing is tricky with dim objects. If you have a computerized scope, we recommend slewing to a nearby bright star and doing your initial focus there. Medium brightness stars provide better focusing information than really bright stars because the latter produce a huge washed out disk. Once your focus is approximate, we highly recommend using a Hartmann mask or diffraction spikes to refine it even further. Remember to refine your focus

Table 4.5. The most common programs used for capturing images from long-exposure modified webcams.

Maxim CCD
AstroArt
K3CCDTools for Windows
Iris for Windows
Astro-snap for Windows
AstroVideo for Windows
Keith's AstroImager for Mac
Equinox for Mac
Qastrocam for Linux
Many more....

Figure 4.7. M104, Sombrero Galaxy. Captured with an SC1.5 Vesta by Keith Wiley with a Meade 8″ f/6.3 LX200 and Mogg 0.6 focal reducer. 47 × 90 sec. Keith's Image Stacker.

throughout an imaging session as changes in temperature and the angle of the OTA tend to throw the focus off slowly.

Once you are focused, you are pretty much ready to go. Different programs have slightly different interfaces, but you basically need to find the exposure time that suits your purposes and start capturing images. As we will see, there is a balance to be struck between capturing a lot of short exposure images and a few much longer ones. The balance will differ between different subjects but as a rule of thumb collecting lots of short exposures will give high image quality at the expense of detecting faint objects and vice versa.

So, how many images is enough? Webcams are not precision instruments and suffer from a large degree of random noise and detrimental artifacts that result from such causes as warm temperatures, electrical interference, and the on-chip amp. These problems can be reduced in a variety of ways, but ultimately no single raw image will be very impressive.

Stacking multiple images offers several advantages, one of which is reducing noise. Additionally, webcams do not have a very good dynamic range because, by the time they transfer their images to the computer, the images are reduced to only 8 bits. This means that objects with a wide dynamic range (such as most nebulae and galaxies) will be unobtainable in a single image. Bright parts of an object will saturate before you have recorded any discernable signal from dim objects. Stacking can reduce this problem by increasing the dynamic range.

Figure 4.8. NGC7635, Bubble Nebula. Captured with an SC Vesta by Etienne Bonduelle with a Meade LX-90 scope and a 0.33× focal reducer. 91 × 60 sec. Astrosnap, Iris, Paint Shop Pro 7 and Neat Image.

You will get smaller and smaller improvements for greater and greater amounts of stacking. However, when possible, a huge number of images is still the best. With deep-sky imaging, you might take images with exposure times ranging between 15 seconds and 3 minutes. Clearly, you cannot capture more than a certain number of frames while the object is well placed in the sky. In short, the more frames you have the patience to collect, the better your final results will be.

Before you take your scope down for the evening, it is important to take some dark frames as well. A dark frame is an image taken in complete darkness, say with the telescope cover on, or simply with a tight cover over the end of the camera-telescope adapter after removing the camera from the scope. The dark frames should match both the exposure time and the temperature of the actual frames. Dark frames suffer from the same maladies of noise as do actual images, so it is a good idea to take several dark frames and stack them to produce a single final dark frame that can be applied to the raw frames later. In addition to a dark frame, it is a good idea to take a flat field frame. This needn't be done each session. Once will be enough. A flat field is most easily captured against a twilight sky and represents an even illumination of the camera's CCD. Flat fields can be

Figure 4.9. M1, Crab Nebula. Captured with an SC3 Vesta by Carsten Arnholm with a C8 scope and a Mogg 0.6× focal reducer. 55 × 40 sec. K3CCDTools, Registax2 and Photoshop.

used to counteract the effects of vignetting (a dimming of the edges of a image) and artifacts caused by dust on the CCD, though it could be argued that keeping the CCD clean is easier than processing out the dust artifacts!

Post-processing Techniques

After you capture a series of long-exposure frames to your computer's hard drive and after you go inside to warm up, it is time to begin the serious task of post-processing your frames with powerful image processing software. Don't despair. The thing to realize is that the post-processing stage represents half the effort and half the fun! This hobby is different from other forms of astrophotography in that it is highly dependent on powerful image processing techniques, and using those techniques to produce a beautiful final image is a large part of the satisfaction.

The most common methods of post-processing, in the order in which they are generally applied, are:

- Stacking of multiple dark frames to a single dark frame.
- Stacking of multiple flat frames to a single flat frame.
- Selection of best frames.
- Subtracting the dark frame from each raw frame.
- Dividing each frame by the flat field.
- Aligning the frames to one another.
- Stacking the frames.
- Sharpening the stack.
- De-noising the stack.
- Level-adjusting the stack.

Many of these steps require software that is specifically designed for processing astrophotos, most of which are available free on the Web. Table 4.6 shows a list of the most popular post-processing programs.

Start by producing a single dark frame that you will use for all of your raw frames. Do this by stacking all the dark frames you took and saving the result. Each program has a slightly different interface, but the basic task is the same.

After dark frame subtraction on each raw frame, divide each result by the flat field frame. At this point you have removed most of the thermal noise and accounted for most of the unevenness in the sensitivity of the CCD. This is what dark frame subtraction and flat field division do. You are left with a series of frames that represent the true theoretical image the camera received, plus a rather large amount of generally random noise. The noise makes the image look fairly undesirable, but stacking will help.

In many cases, a number of your raw frames will be hopelessly degraded. There are a variety of possible degradations, including polar alignment error, periodic drive error, planes and satellites flying through the exposure, blurred images caused by wind gusts blowing the mount, bad seeing and bad focus. Simply jettison any bad frames. In deep-sky imaging this is fairly easy because there aren't many frames to sift through.

You must carefully align all the frames to each other. In most programs you do this by designating one particularly good frame and aligning all the other frames to that frame. There are many ways in which alignment might be required, such as translation, rotation, and stretching. In practice, translational alignment is probably the only method you need to perform if your scope is polar-aligned.

Table 4.6. The most popular post-processing programs used for processing sets of images captured from long-exposure modified webcams.

Registax for Windows
K3CCDTools for Windows
iMerge for Windows
Adobe Photoshop for Windows and Mac
Keith's Image Stacker for Mac

Figure 4.10. IC2177, Seagull Nebula. Captured with an SC3 ToUcam by Jim Hommes with an ST f/2.5 scope. 6 × 180 sec H-alpha, 16 × 8 sec IRB, 12 × 12 sec RGB. 4 frame mosaic. K3CDTools, Astroart and Photoshop.

At this stage you generate the stack, which is a single operation without much effort on your part. How the image is represented on the screen depends on which program you are using, but you can be sure they are all performing some impressive tricks to squeeze the stack's large dynamic range into the mere 8-bit depth common to most image formats and computer screens. Nevertheless, the actual stack, as stored in the computer's memory, contains the full dynamic range of the combined data.

Stacking performs two feats at once. It increases both the signal-to-noise ratio (S/N) of the final image and its dynamic range (see Figure 4.11). It is the first of these, the S/N increase, that makes the stack look smoother and less grainy than the individual raw frames. The degree to which stacking accomplishes this feat scales with the square root of the number of frames you stack. So when you stack four frames, your S/N goes up by a factor of two. Notice, however, that to get another factor of two (so a factor of four overall by comparison to the raw

Figure 4.11. M42/M43, Orion Nebula. Captured with an SC1.5 Vesta by Keith Wiley with a Meade 8″ f/6.3 LX200. Exposures ranging from 5 to 160 sec, stacked between 18 and 80 deep. 25 frame mosaic. Keith's Image Stacker and Photoshop 7.

frames), you need not eight frames, as you might initially expect (because that is twice the previous stack), but sixteen frames. This explains why, as more and more images are added, the improvement becomes less discernible. For this reason, you should not get too concerned with whether you have 30 or 34 frames, or whether you have 100 or 120 frames. The results will vary by almost unnoticeable amounts in such cases. Once you have a stack you can do a lot of neat things to further refine the final result. Sharpening is generally more useful on planetary stacks, but it can be useful in deep-sky ones as well. There are a few different sharpening techniques, most of which are available in the stacking programs mentioned earlier.

To clean up the noise of the final stack further, there are some advanced methods of noise reduction, most of which use wavelets. It is always preferable to first reduce the noise through stacking, as stacking is the only method of noise reduction that is guaranteed to approximate the true-recorded signal. Other denoising techniques, such as wavelet-shrinkage and expectation-maximization, make assumptions and approximations in their efforts to reduce noise.

Level adjustment is important. The stack will contain information from the bright cores of nebulae and galaxies as well as dim information from the perimeter of these objects. Proper level adjustment is crucial to bringing up the dim areas without blowing out the bright areas.

Figure 4.12. NGC2244, Rosette Nebula. Captured with an SC3 ToUcam by Jim Hommes with an ST f/2.5 scope. 10 × 90 H-alpha, 18 × 4 IRB, 16 × 4 RGB. 4 frame mosaic. K3CCDTools, Astroart and Photoshop.

Conclusions

Webcam astrophotography is only a few years old, and the progress that has been made so far is astounding. We invite you to join the online group QCUIAG. You can learn how to get involved in this hobby yourself for a fraction of the price of professional astrophotography equipment.

Relevant Web Pages

Jim Hommes – http://jthommes.com/Astro/

Carsten A. Arnholm – http://arnholm.org/

Etienne Bonduelle – http://www.astrosurf.com/astrobond/

Ashley Roeckelein – http://astro.ai-software.com/

CHAPTER FIVE

Deep-Sky Imaging with a Digital SLR

Johannes Schedler

Introduction

Digital cameras (digicams) are now dominating the sales for general purpose photography. In the past 2 years, besides the point-and-shoot cameras, digital SLRs have been replacing film SLRs, and at competitive prices. All types of digicams have begun to be used for shooting the Moon, the planets and the Sun (with appropriate filter) – and with great success. However, these objects can all be imaged with short exposures similar to those for daytime imaging. The use of digicams for deep-sky imaging has been very limited due to the fact that these cameras have uncooled CCD/CMOS chips. This design is very good for standard shooting conditions under available daylight but has proved very problematical for real long-time exposures. Most of the point-and-shoot cameras are limited to exposure times between 4 and 30 seconds. Even the presence of a "bulb" mode does not help much as the noise overwhelms the signal for most cameras. So only bright objects could be imaged successfully at that time.

The Arrival of (Affordable) Digital SLRs

In June 2002 Canon released its D60 D-SLR model, one of the first 6.3-megapixel cameras. Initial tests published for this camera showed much lower noise in long exposures than all previous models. I acquired one of the first available models.

Figure 5.1. Johannes Schedler in his observatory with his 4-inch TMB "Apo" and Celestron C11 Schmidt-Cassegrain.

Early trials soon showed a dramatic improvement in reducing noise for long exposures. For the first time, exposures of more than 60 seconds could be used and even at summer temperatures. What could be achieved with these new D-SLR cameras at that time was reviewed in the *Sky & Telescope* October 2002 and June 2004 issues, with the D-SLR results featured in the "Gallery" section. For the following year, I used the D60 extensively to image many of the brighter deep-sky objects, most of them at prime focus of my 4-inch TMB refractor and Celestron C11 (at f/6) telescopes (see Figure 5.1).

In June 2003 the Canon D60 was replaced with the new model 10D (see Figure 5.2). Although basically the same chip design, there were several key improvements introduced with it, namely:

- significantly better noise performance for long exposures for all ISO ratings;
- nearly complete elimination of the red amplifier glow on the right side of the image;
- higher ISO settings up to 1600 (3200 not really usable);
- higher review magnification of up to 10× for analyzing images via the integral display;
- improved remote capture software allowing multiple interval exposures up to a maximum of 30 sec each (but not in bulb mode!).

See Figure 5.3 for the noise comparison of the 10D compared to the D60 showing a 300-sec ISO800 dark at 22°C, cropped from the center to the right corner.

Figure 5.2. Camera Canon D60 and Canon 10D.

The new cheaper Canon Rebel (300D in Europe), built with a plastic casing, shows similar results for deep sky applications but two major drawbacks have to be noted:

1. No mirror lockup is possible, but a firmware update, available from the Canon homepage, enables a mirror lockup.
2. No direct connection to remote interval timer is possible – the plug for the timer TC80N3 must be exchanged to enable its use.

The SLR-type camera body provides an easy coupling to any telescope at prime focus. This connection is usually by means of a T-ring that attaches to typical telescope accessories. The large chips of the new D-SLRs offer better quality and wider coverage of the telescopic field than is possible with the smaller chips of most astronomical cameras. Field flatteners/reducers are very helpful in achieving the desired image quality into the frame corners.

What are the biggest advantages of such cameras compared to film SLRs? A clear advantage, especially for photographers with little experience, is the

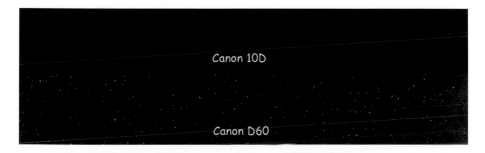

Figure 5.3. Comparison image of the noise performance – Canon 10D vs. D60.

Figure 5.4. Star trails, imaged on a tripod with the Canon 10D and a 20mm lens.

immediate review of the image by getting the result quickly after the exposure. This allows problems related to focus, composition and sky background level to be resolved during the imaging session itself. A second advantage is the relatively low noise and high resolution compared to a typical slide or negative color film. New cameras like the 10D not only save the images in jpg mode but also in raw mode that utilizes the internal 12-bit-per-channel format. This can further be converted to 16-bit-per-channel tiff images. This is a big advantage for the wide contrast

Table 5.1. Overview on current D-SLR camera models (compared in early 2005 for DSO capability)

Model	Effective Pixel	Deep-Sky Capability	Street Price Approx. in $ (€)	Comment
Canon 1DS	11 MP	+++	7000	Top model, 24 × 36mm chip. Replaced with MKII model with 16.7 MP
Canon 10D	6.3 MP	+++	1500	Best all-round performance. Replaced by 20D model with 8.2 MP
Canon Rebel (300D)	6.3 MP	+++	1000	Best price/performance ratio, limited functions
Fuji S2 Pro	6.3 MP	++	1800	Good all-round performance. Replaced by S3 model with 6 MP (×2)
Kodak DSC Pro14	13 MP	+	6000	Bad noise behavior at higher ISO
Nikon D100	6.3 MP	++	1500	Good all-round performance
Nikon D2H	4.1 MP	++	3000	Good all-round performance
Nikon D70	6.3 MP	++(+)	1000	Low noise, but amplifier glow, best alternative to Canon
Olympus E1	5 MP	++	2200	Good all-round performance, some noise
Pentax *ist D	6.1 MP	++	1350	Good all-round performance, some noise. DS model similar at lower cost
Sigma SD9	3.5 MP per color	–	1100	Foveon X3 sensor, weak low-light performance, no bulb mode
Sigma SD10	3.5 MP per color	+	1600	Foveon X3 sensor, improved performance at ISO400

Note: The noise behavior at long exposures is not comparable to daytime behavior. All the listed cameras are excellent for use during daytime.

range that has to be accommodated with many deep-sky images, e.g., faint nebula structures embedded in bright foreground stars. Compared to 35mm film, when using lenses, the focal length has to be multiplied by a factor of 1.6 to get the equivalent field coverage. See Table 5.1 for an overview of current DSLR cameras.

As expected, the noise behavior very much depends on the ambient temperature. As is the case with cooled CCDs, the sensors of digicams are subject to dark current, which doubles for every 6°C increase in temperature. However, some of the dark noise is compensated for by the internal calculations of the CMOS chip. During the warm season a well-matched dark frame subtraction is essential to get the best results out of the raw images. During wintertime, even at high ISO ratings, the dark current and noise are reduced to a minimum so that a dark frame subtraction is not as essential, since the residual noise is basically of a random structure.

The graph (Figure 5.5) illustrates the standard deviation for 300-sec darks at two different temperatures (22°C vs. −4°C). These darks have been taken in raw mode, converted to 16-bit-per-channel images by ImagesPlus and examined in Astroart.

The values in Figure 5.5 demonstrate that there is little advantage in using higher ISO ratings for achieving maximum signal to noise (S/N). For my typical semi-rural sky conditions, I use ISO 200–800 for unfiltered images and ISO 400–1600 for narrowband filtered images. For long exposures, I use 5 minute

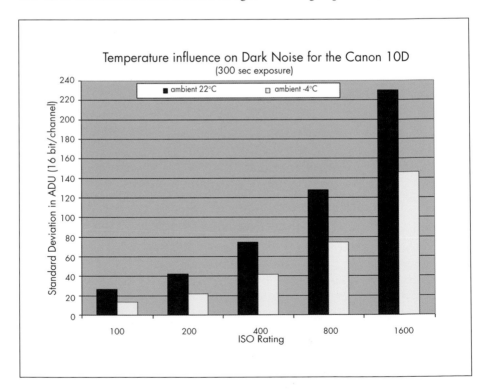

Figure 5.5. Standard deviation of darks at 22°C vs. −4°C.

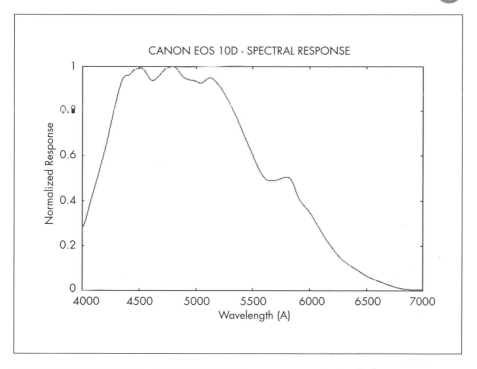

CANON EOS 10D - SPECTRAL RESPONSE

Figure 5.6. Relative spectral response of the Canon 10D. Note the low response at the important 6563Å H-alpha emission line.

exposure for raw frames with the white balance set to "sunny." For shorter exposures (less than 2 minutes), the lower ISO settings prove the best.

There is one major disadvantage for the D-SLRs compared to film imaging: the limited sensitivity in the far red where the H-alpha emission line at 6563 Å reveals most detailed structures in the majority of emission nebulae (see Figure 5.6). The spectral sensitivity investigations from Christian Buil on the 10D are showing that the normalized sensitivity falls off from 100% in the green to approximately 6% at 6563 Å. Considering a 40% average absolute quantum efficiency for ABG chips, the absolute quantum efficiency at 6563Å will be near 2%.

The integral IR blocking filter of the 10D reduces the far red end of the visible spectra and therefore is the main reason for the poor red response. Preliminary tests on D-SLR cameras with the IR filter removed showed a significant increase of the red response. However, carrying out this modification cannot be recommended because the filter is an important part in the optical path. Removal will make the auto-focus useless and the color balance will be severely disturbed for daylight imaging. Hutech is now marketing a modified Canon Rebel camera, with an exchanged IR-cut filter optimized for dedicated astro-imaging purposes, for approximately $1500. Because the Bayer color filter pattern used on the chip contains 50% green pixels and 25% for both blue and red pixels, the expected absolute sensitivity without the IR-cut filter will not exceed 10% at 6563 Å over the total chip area, even for the modified cameras.

Figure 5.7. The Crescent Nebula NGC6888 imaged with a Canon 10D and Celestron C11.

The biggest disadvantage of CCDs compared to film, until now, has been the much smaller chip size compared to the film format (typical 36 × 24mm). This has changed, however, in the past two years. Like the previous D60, the Canon 10D uses the same 22.7 × 15.1mm 6.3 megapixel CMOS sensor (3072 × 2048 final image size). This corresponds to the size of APS film format. The big chip size compensates for the reduced red sensitivity and the noise at high ambient temperatures.

The already mentioned limitations of the D-SLRs are valid again when comparing them to dedicated astronomical CCD cameras. These typically monochrome cameras are cooled to a constant low temperature of approximately 30°C below ambient. They are optimized for quantum efficiency and they can reach between 50% and 80% over the visible wavelengths. In a monochrome CCD camera every imaging pixel contributes for maximum resolution as no interpolation between the different colors filters has to be performed. CCD astro cameras show real 16-bit-per-pixel resolution and are close to ideal, even under heavily light-polluted skies. My opinion is that for the foreseeable future, dedicated astronomical cameras will have significant advantages for narrowband imaging and for reaching the faintest structures. Additional features that D-SLRs cannot provide are binning, subframe readout, focus support routines, integrated guiding and automated image acquisition, as well as links to other astronomical programs and hardware.

What features should one look for in a digital SLR for astrophotography? The camera must have a bulb mode and connectivity to a remote interval timer, either hardware like the TC80N3 for the Canon D-SLRs or software based via the PC connection for long controlled exposures. One should select a model with low noise design, as shown in Figure 5.3. I would recommend the current standard line of D-SLRs; the top-end models will not repay the high investment, as technical progress will push forward the features like chip size, transfer speed and sensitivity in high speed.

The model should be able to use existing high-quality lenses. Focal lengths between 20 and 300 mm are able to frame most of the possible deep-sky targets. Figure 5.8 shows my typical setup for deep-sky imaging using the C11 with a reducer for f/6, mirror lock and helical fine focuser for the D-SLR. A 4-inch TMB refractor at f/12.5 is used as a guide scope with the MX7C autoguiding. The setup is also used with the telescope roles reversed. The D-SLR is permanently connected to the PC to control focusing, while the timer TC80N3 automatically takes the exposure sequence.

Imaging and Processing

Focusing is the first critical step for any imaging session. To avoid focus shifting caused by the main mirror of SCTs, the mirror should be locked. First I calibrate my go-to mount and use a bright star to focus the D-SLR. Visual focusing is easier when using the 2.5× angle viewer. A Hartmann mask or two parallel tapes across the lens improve the judging of the best focus. After determining the best focus it is essential to prove the focus by taking a test image of say 30 seconds and then checking it on the PC. At the optimum focus, sharp spikes and interference patterns on bright stars can be seen from the tapes. The freeware *DSLRFOCUS* helps with this procedure. Then the scope can be slewed to the desired object and the imaging can begin. Fast lenses should be stopped down one or two steps after focusing. During the night, as the temperature is typically dropping, a refocusing should be considered every one to two hours.

The quality of a deep-sky image proportionally increases with the S/N ratio. Doubling the exposure time improves the S/N by $\sqrt{2} = 1.4$ and so on. Taking many raw images is the key to deep and high-quality images. Find out what maximum exposure time your mount is capable of. My setup with separate guide scope achieves 5- to 10-minute subexposures. If combining say 16 raw images, you should also collect 10 to 16 dark frames under the same conditions. They can be taken automatically with the telescope or lens closed.

For processing large raw files, *ImagesPlus* image processing software is very capable and reliable, able to perform all the necessary steps. First, average the darks and make a master dark of them. Then load the master dark as reference in the calibration setup and calibrate your raw images. If you have vignetting in your optical system, you should also include a flat for calibration. Next step is aligning the calibrated images, easily performed by "Image File Operations/ Align File/ Translate, Scale, Rotate" by marking reference stars. When using Canon raw files, convert them to 16-bit tiff files first, then align and average them in 16 bit. When using 8-bit jpg files, you should use the "extended add" for combining.

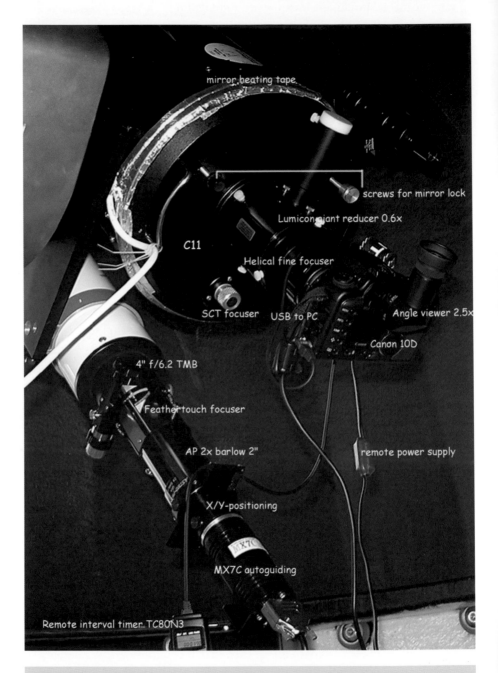

Figure 5.8. Johannes Schedler's imaging setup. In this case the Celestron C11 is being used to image with the Canon 10D and the TMB refractor for autoguiding.

Figure 5.9. The Globular cluster M13 imaged with a Canon 10D and 4-inch APO refractor. Note the faint galaxy recorded in the top left-hand corner.

I have found that combining a large number of jpg-fine images does not show a major disadvantage compared to using the raw mode. "Extended adding" of 16 calibrated 8-bit images in the 16-bit-per-channel space increases the dynamic range of the final image from 8 to 12 bits. This compensates for the limited dynamic range of the 8-bit-per-color jpg files. A second advantage of the "extended adding" is the combining of different exposure times, i.e., for M42 exposures of 20 sec, 60 sec and 300 sec can be added together in one step. This procedure is much easier than difficult masking procedures for the burned-out areas (see Figures 5.10 and 5.13). The calibrated combined images should be saved in 16 bit tiff format, then a DDP processing (i.e., in ImagesPlus) supplies the desired nonlinear stretching. Further processing steps, like removing gradients and adjusting the color balance, I prefer to do in Photoshop. In most

22 sec ISO200 raw   60 sec ISO200 raw   300 sec ISO400 raw   added and processed

Figure 5.10. Adding exposures of different lengths preserves the detail that would otherwise be burnt out in the brightest areas. This image comprises crops of 3 raws from the left to right that have been added in the following way: 2×20 sec, 8×60 sec and 26×5 min. The final combined result is shown in the image crop at right. The telescope was the 4-inch TMB refractor with reducer at f/5.

cases a noise reduction program like *Neatimage* helps in smoothing the final image.

Hints and Tips

Using fast telescopes or lenses with large and flat illuminated fields makes imaging much easier. Try tripod shots for star trail images. Start with piggyback images and short focal lengths. Constellation images are very forgiving of imperfect tracking. M31 with a 50mm lens is an impressive target. With more experience, longer focal lengths can be used. High-quality fast refractors with flat field like the Televue NP101 and imaging dedicated telescopes like the Takahashi Epsilon are best suited for use with D-SLRs. Above 300 mm focal length, an autoguider is highly recommended.

In many situations we have to live with light-polluted skies. Because of the 8-to 12-bit resolution of the D-SLRs, they are more sensitive to light pollution than astro cameras. Filters are a big help in overcoming these effects. The Hutech LPS filter and high-quality UHC filters allow us to reveal faint nebula structures even under magnitude 4.5 skies, while star colors can be preserved. Narrowband filters like the Astronomik 15 nm H-alpha filter need very fast lenses (< f/2.8) and low temperatures for optimum results. The better your raws and the more raws you have, the less effort will be needed to achieve the best final image quality.

As we all have to live with the given sky conditions, I select the objects to fit best with the prevailing conditions:

- Under moon-lit skies, small bright planetary nebulae and globular clusters can be imaged. Narrowband filters allow nebula imaging even with a bright sky background.

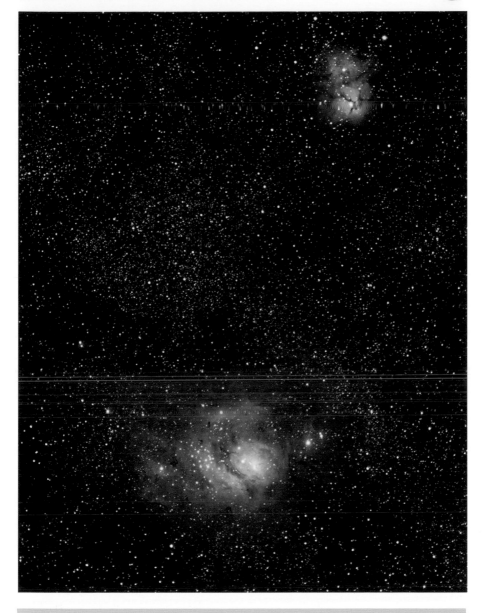

Figure 5.11. Lagoon and Trifid Nebulae, M8 and M20.

- When seeing is unsteady, I use the shorter focal length refractor and lenses for extended nebulae, open clusters and star fields (see Figure 5.11).
- The best dark and steady nights should be reserved for long focal lengths to image faint distant targets like galaxies. If possible, we should try to image difficult targets near the meridian, the point of highest elevation.

Figure 5.12. The Great Andromeda Galaxy, M31.

Conclusion

Modern D-SLRs are now available from $1000, which represents a much lower cost than most of the cooled astro cameras. They are multipurpose cameras usable for both daytime and nighttime imaging. They are well suited for gaining experience in the fascinating world of deep-sky imaging because they can provide stunning results on many of the brighter deep-sky objects and all in a simple one-shot color technique. However, the higher noise (compared to most dedicated astro cameras) must be compensated for by more effort and expertise in the image processing stage. In my opinion, astronomical cameras will always be some steps ahead in terms of sensitivity and versatility, especially for narrowband imaging, but at a much higher investment cost.

Figure 5.13. The Great Orion Nebula M42, M43 and NGC1977.

Useful related links

Yahoo Group DIGITAL_ASTRO:
http://groups.yahoo.com/group/digital_astro/

FAQ–Digicams for Astronomical Use:
http://www.szykman.com/Astro/AstroDigiCamFAQ.html

10D Spectral Sensitivity Charts (French):
http://astrosurf.com/buil/us/digit/spectra.htm

S/N comparisons by Roger Clark:
http://clarkvision.com/astro/canon-10d-signal-to-noise/

ImagesPlus Image Processing Software: http://www.mlunsold.com/

Al Kelly's guide to Acquiring and Processing:
http://www.ghg.net/akelly/procccd.htm

Photoshop for Astrophotographers by Jerry Lodriguss:
http://www.astropix.com/PFA/INTRO.HTM

The new CCD Astronomy by Ron Wodaski:
http://www.newastro.com/newastro/default.asp

DSLRFOCUS for Canon DSLRs:
http://www.dslrfocus.com

Section 2

Getting Serious

CHAPTER SIX

IRIS: Astronomical Image-Processing Software

Christian Buil

Introduction

The genesis of the IRIS software is a long story. Its ancestor was born in the middle of the 1980s, when the first experimental CCD cameras for amateurs were appearing. It was written for the Apple III computer, in assembly language, and had the capacity to process images with a size up to 128×128 pixels! As more and more powerful cameras arrived on the scene, there was an increasing need for more image-processing power. IRIS steadily grew and took the form of what it is today: running on the Windows platform and able to process images of several thousand pixels. IRIS is freeware and can be downloaded from the Internet at the following address:

http://www.astrosurf.com/buil/iris/iris.htm.

This chapter is a quick overview of the philosophy and possibilities of IRIS. First and foremost, IRIS software is primarily orientated to the processing and scientific analysis of images. Final presentation functions are generally few in number as these are well covered with common graphics software. However, these processes are generally nonlinear, which destroys the photometry of the image content, so it is important just to use them for final touching up. Nevertheless, in the latest release, IRIS does have some processing and display functions for true color images, as the use (and importance) of new digital cameras is growing in astronomical observation. We will come back to this topic at the end of the chapter.

Figure 6.1. Christian Buil and the T60 telescope at the Pic du Midi Observatory.

Calibration of Images

The sequence of operations on an electronic image, just after the acquisition and before it can be effectively analyzed, is very well documented. The first requirement is offset subtraction, which is equivalent to determining the zero point for each pixel (termed *bias* or *offset*). The second is the removal of the thermal signal produced in the sensor itself (termed *dark current* or *dark-frame*). The last is achieving pixel uniformity across the entire image field (termed *flat-field*). These operations are all elementary, but it is critical to carry them out optimally so as not to waste the precious information we have patiently gathered at the telescope. They are also very repetitive as they have to be applied to every electronic image. Let us see how IRIS solves these problems.

From the very beginning, IRIS has had a command-line user interface. When mouse clicks and pop-up windows did not exist, it was the only way to give instructions to a program. Even though IRIS has now moved to a Windows-oriented style, where basic commands are available in drop-down menus, IRIS still makes available a "console mode" where we can enter and edit commands (see Figure 6.2). This method of controlling software might be regarded as old-fashioned (and perhaps obsolete) but in reality it is more flexible and powerful than a mouse-click navigation style, where windows, menus and dialog boxes can completely cover the screen. Consider what professional astronomers use. They almost exclusively use command-line scripts in such standard professional software as IRAF or MIDAS. What is best for them is surely good enough for us! A common shortcut, when using a command-line interface, is not to type new commands in full but to edit previous ones. So when using a similar series of commands, only the first has to be typed in full – subsequent commands are edits of the previous one. This makes processing a sequence of images quick and straightforward. By default, the images are initially loaded and stored in a working directory. This directory is set up from a drop-down dialog box, as are all the general program parameters. As might be expected, the setup is automatically saved for the next usage.

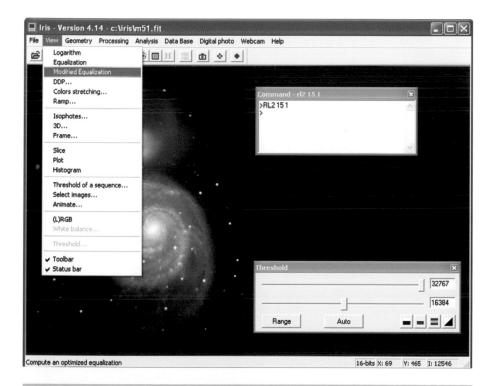

Figure 6.2. Screen shot of IRIS in action. Richardson-Lucy Deconvolution has been carried out via the command line (RL2 15 1) and the result viewed with Modified Equalization selected from a drop-down menu.

The second stage of processing makes use of the operation SUB (i.e., subtract IMAGE1 – IMAGE2). The two parameters are the operand file names and a third is a constant added to each pixel after subtraction. For example, for a null value constant the full command is:

SUB IMAGE1 IMAGE2 0

The result of the subtraction is in memory and is displayed on the screen. To keep the result it needs to be saved on the hard disk with the SAVE command. IRIS is also able to process a sequence of images. As an example, to subtract an offset image from the five images RAW1, RAW2, RAW3, RAW4, RAW5, only one command is used:

SUB2 RAW OFFSET RESULT 0 5

The first parameter is the generic name of the sequence to be processed, i.e., the name of the image file without the index number. IRIS adds the index itself to find the correct file on the hard disk. The second parameter is the offset image name to be subtracted from the sequence of raw images. The third parameter is the generic name of the resulting image sequence; again IRIS will add the index. The fourth parameter is a constant value to be added to each pixel of the resulting images. Finally, the fifth parameter is the number of images to be processed in the sequence. The command SUB2 therefore executes all the operations: RAW1–OFFSET=RESULT1, RAW2–OFFSET=RESULT2, etc.

This syntax is very common in IRIS. The "2" appended to a command (e.g., SUB2) generally denotes it is for processing a sequence. Similarly, for dividing by a flat-field:

DIV2 IMAG FLAT IMAG 3

This performs the operations: IMAG1/FLAT=IMAG1, IMAG2/FLAT=IMAG2, IMAG3/FLAT=IMAG3. Here, the output sequence has the same name as the input sequence, which saves space on the hard disk but makes the operation non-reversible: Use with care! The command DIV2 performs a little bit more than a simple division as it normalizes the median level of the input images, which then preserves the mean level of the processed images.

IRIS includes many complex commands (which are also available in drop-down menus). As an example, to compute the median sum of an image sequence, it is sufficient to enter from the command line:

SMEDIAN IMAG 15

SMEDIAN (literally *stack median*) computes the median pixel from the pixels in an image sequence of 15 images with a generic name "IMAG." The result is displayed on the screen; again, it can be saved in a file with the SAVE command.

IRIS includes more than 200 commands. When it is considered that many of them have multiple suboptions, the total number of available functions, from the console, is estimated to be close to 5000! Their description is beyond the scope of this chapter. For an exhaustive list of IRIS functions it is better to refer to the Web page: http://astrosurf.com/buil/us/iris/iris8.htm.

Let us go a bit deeper into the key steps of the astronomical imaging process by first considering how best to remove the thermal component from the raw image. The thermal signal, also called the dark signal or current, is an artifact due to the spontaneous generation of electrical charges within the sensor itself, triggered by temperature. Despite considerable progress over the last decade in sensor technology (MPP technology, for instance) the thermal signal is still significant in deep-sky imaging, where the exposure duration is generally quite long. In some circumstances images can be taken with a noncooled device, which is the case with a digital camera. The issue of the elimination of the dark signal is tricky because it is not usually a constant, as temperature could have varied from one exposure to the next. Under these circumstances, if a unique dark image, acquired in darkness with the same exposure length, is subtracted from all the images, the result will not be reliable. IRIS uses an optimal function of the thermal signal, which compensates for these variations in producing the calibrated image. The basic idea is to multiply the dark image by a value, which is computed to minimize the noise in the resulting image. The constant value c is given by the formula:

$$ c = \frac{n \cdot \sum S_{i,j} \cdot D_{i,j} - \sum S_{i,j} \cdot \sum D_{i,j}}{n \cdot \sum D_{i,j}^2 - \left[\sum D_{i,j}\right]^2}. $$

With $S_{i,j}$ the pixel intensity at i,j coordinates on the processed image, $D_{i,j}$ is the corresponding intensity in the dark image and n is the number of pixels concerned. The user selects the area where the computation is to be carried out by simply dragging a box with the mouse. As an example, if the raw image is already loaded in memory, the command OPT DARK computes the optimal coefficient using the master dark-frame image (DARK) stored on the hard disk, multiplies each pixel in the image by this constant and subtracts the result from the raw image. As you may have guessed, the related command OPT2 is available to automatically process an image sequence.

Alignment of Images

Most of the time, long exposures in astronomical imaging are achieved by a series of shorter ones. These are then added digitally during image processing. This requires that each image be aligned, before the addition, to cater for the telescope field of view shifting between exposures. This can be a random occurrence or deliberately achieved (see further in this chapter). It is a classical problem in image processing. For instance, in a sequence of n images, the shift of the second image is interpolated so it can be correctly superimposed onto the first image. The same for the third and so on up to the nth image. The interpolation method can be bilinear or it can use the spline function in IRIS. The only true difficulty is to find common details between all images so that the software builds on them to compute the coefficient of the geometrical transformation. The easiest case, and the most common one, is that of a deep-sky sequence where a single star can be selected. Here, a simple command like REGISTER can recenter the images by computing the translation values, in the two axes, to a fraction of a pixel. As an example:

REGISTER IN OUT 21

This recenters the image sequence IN1, IN2, ... IN21 and produces a new sequence OUT1, OUT2, ... OUT21.

There is a more complex case, where the image field contains many stars or where the geometrical transformation comprises translations, rotations and scale distortion. Thus, for the deep-sky, the command COREGISTER is more powerful, e.g.:

COREGISTER IMAGE1 IMAGE2

This automatically detects the common stars in the images specified and calculates an accurate registration of IMAGE2 on IMAGE1, by applying the required translations, rotations and scale change. This type of function is especially useful where images have been acquired using different telescopes. The power of the COREGISTER command is available with several variations, especially for the totally automatic preprocessing of an image sequence. The algorithm uses the detection of a triangle by tracing lines between a star triplet.

Planetary image registration is also available in IRIS. One of the methods to achieve it (command PREGISTER) is by computing the intercorrelation function between two images. The algorithm uses the Fourier transform. Another technique, more simple but very efficient on images having a planetary disk of high contrast, consists of adjusting, by the mean-square method, a circle to fit the limb. The resulting accurate knowledge of the center of the disk in each image is then used to compute the shift value to a fraction of a pixel (command CREGISTER).

The addition of an image sequence can be just a simple arithmetic sum of all the pixel intensities (ADD2 command). IRIS has more sophisticated options: to either improve the signal-to-noise ratio or to increase the spatial resolution. The advantage of dividing a long exposure into a sum of smaller ones is to be able to perform statistical processing that eliminates the pixels having a spurious intensity, for example, if the pixel intensity is altered by a cosmic ray. The standard median composite function applied to an image sequence is a simple and robust rejection method for deviating pixels. However, the drawback is the degradation of the signal-to-noise ratio by a factor 4/pi compared to a simple addition. The command COMPOSIT uses a rejection technique with better characteristics, known as *sigma-clipping*. For a given pixel in the sequence, the software eliminates those values that differ greatly from the standard deviation of the pixel intensity distribution. This process can be used iteratively for maximum efficiency. The signal-to-noise ratio is preserved with this technique. Sigma-clipping should be mastered by anyone who wants to get the best from their images.

Improving Spatial Resolution

If, between each exposure of a sequence, the telescope has slightly moved, it is possible to improve the spatial resolution during image stacking. The DRIZZLE command uses an algorithm, which has been used with great success on images from the Hubble Space Telescope (see, for example, http://www.stsci.edu/~fruchter/dipher/dipher.html). Basically, when the images are shifted and recentered, the final image is assembled with a pixel grid having a smaller pitch than the original. The resolution gain is real only if the original images are undersampled, i.e., when the telescope focal length is too short for the pixel size. One may consider the system is undersampled when the FWHM is smaller than 2 pixels. Images acquired with excellent apochromatic refractors or with very good photographic lenses often have this characteristic. A sequence of at least ten images should be available and the shift needs to be random in the two axes. The acquisition method for producing these deliberately is called *dithering*. Results using the DRIZZLE command can be spectacular with amateur instruments. This is especially true when the focal length is small enough and the seeing (atmospheric turbulence) spreads out, a little bit, the image detail compared to the pixel size. Figures 6.3 and 6.4 show the results, on the same image sequence, of using a simple addition compared to one using the DRIZZLE command.

Analysis Tools

The IRIS software also contains many analysis tools. These cover photometry, astrometry, polarimetry and spectroscopy. Most of these functions are available through dialog boxes, as many parameters have to be entered. For instance, for photometry, it is possible to choose the aperture circle method, with an

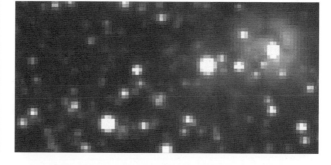

Figure 6.3.
Composite of several images combined using simple addition.

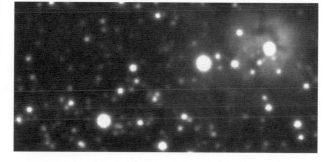

Figure 6.4.
Composite of several images combined using the DRIZZLE command.

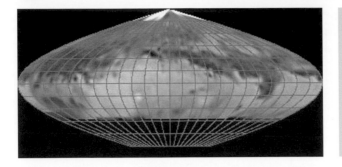

Figure 6.5.
Mars – sinusoidal
(Sanson-Flamsteed)
projection map.

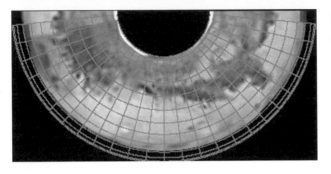

Figure 6.6. Mars –
polar conic (Albers)
projection map.

adjustable value for the circle measuring the sky background, or a mathematical function adjusting the star point-spread function to determine its intensity. In the astrometry domain, a database of stars on a CDROM can be used. Figures 6.5 and 6.6 are examples of output using the planetary cartography module.

The module for spectral analysis is especially powerful. It corrects for distortion and extracts the spectral profile, in an optimized way, using a specific algorithm that maximizes the signal-to-noise ratio (see Figures 6.7 and 6.8).

Digital SLR Image Processing

The latest developments in IRIS software have been targeted at digital SLR camera usage. The widespread development of detectors in large sizes, with high performance and low cost, is a direct result of the fierce competition in the digital

Figure 6.7. A two-dimensional image of 59 Cygnii 's spectrum (LHIRES2 spectrograph + 60 cm telescope of the Pic du Midi observatory – R = 20.000).

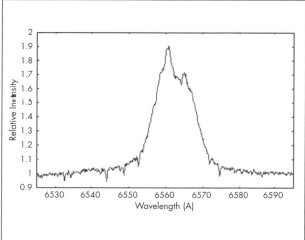

Figure 6.8. The spectral profile extracted from the 2D image in Figure 6.7.

SLR camera marketplace. Digital SLR cameras are simple to employ at the focal plane of our telescope. CCD cameras, optimized for astronomy, are still the best for studies requiring maximum performance, such as deep-sky surveys, but digital cameras have other advantages. These advantages include a single unit, compact size, optical viewfinder and a high-performance, film-sized, color CCD/CMOS sensor. In addition, their cost is falling almost monthly, already becoming much much cheaper than a dedicated CCD camera. The large size of the sensor used can be quite decisive for some scientific studies like nova detection. Overall, digital cameras are not in competition with CCD cameras but rather are complementary.

The image coming out of such digital SLR cameras results from the simultaneous recording of the three-color planes, i.e., red, green and blue. This allows the final restoration of the color image but requires specific processing. To get such results, a tiny color filter covers each pixel in a specific pattern (see Chapter 1). As each color pixel is spatially separated, when combining them, an interpolation algorithm has to be used to recover the final colored image in full resolution. IRIS has alternatives for this but in deep-sky imaging, where noise is predominant, avoiding aggressive algorithms designed to obtain the most resolved image is recommended. A more efficient solution, regarding signal-to-noise ratio optimization that also minimizes artifacts, is usually a simple linear combination of the colored pixels.

The RAW format is virtually mandatory to get the best out of a digital camera. In this format, the camera records the image as it comes straight off the sensor (unfortunately some exceptions exist, like the Nikon D70 where a median filter is systematically applied to the RAW image to eliminate the hot spots, which is nonsense for a RAW image). Images compressed to JPEG format lose precious information in the compression process, which is undesirable for astronomical usage. Every manufacturer has its own RAW format, which greatly complicates the software designer's life! IRIS is able to convert the proprietary RAW format into a standard FITS image for most cameras from the leading manufacturers.

The preprocessing routines applied are then completely identical to that of classical CCD images. Only the specification of the colored filter matrix has to be taken into account and leads to some algorithm adaptation, especially for the flat-field correction or for the white balance function (a solar-type star or a white illuminated screen has to be used).

Summary

In conclusion, we should remember that the optimal display of an image is an important part of image processing if we are to extract the most hidden information. For this, IRIS has a lot of intensity transposition tables, from a basic logarithm up to advanced transformations like *arcsine-hyperbolic,* which is very efficient on color images. Animation functions can also be found and are very useful for showing the motion of an asteroid or comet across a star field.

It must be admitted that the first contact with IRIS, especially for the beginner, is likely to be off-putting. It is not initially simple to use, but the power of the software will become apparent over time. For getting up to speed rapidly, I recommend studying the numerous illustrated examples available on the Web site. These will allow you to enter the IRIS astronomical processing universe.

CHAPTER SEVEN

High-Resolution Imaging of the Planets

Damian A. Peach

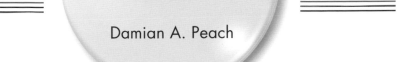

Introduction

I have been an active amateur astronomer for more than 15 years and, ever since my first steps into this immensely rewarding science, observing the finest details on the planets has been enormously fascinating to me. To be able to monitor, in detail, the weather of other worlds from one's own backyard has to be one of the most exiting areas of the science. The planets are bright and easily observable to astronomers of every level of experience, wherever they are located. They provide the amateur with a real opportunity to contribute to our understanding of the weather systems of Earth's nearest neighbors.

The Observatory and Equipment

My observatory has varied throughout my years as an astronomer and currently consists of gardens at the back and front of my home in rural South Buckinghamshire, UK (51° N, 0.5° W), though I have observed from other parts of the southern UK and for long periods in the Canary Islands. My primary telescopes employed for high-resolution imaging were (until recently):

1. Celestron 11-inch (28 cm) Schmidt-Cassegrain Telescope: This is the primary telescope used for my planetary imaging programs. It was acquired new in May 2002 and is mounted on a G11 German Equatorial mount. The telescope is of high quality optically and presents superb views of the planets up to greater

Figure 7.1. An image of Saturn obtained by the author on December 16, 2003. This view of Saturn close to opposition, obtained under excellent atmospheric seeing conditions, reveals a level of detail greatly exceeding the finest Earth-based photographs from professional telescopes taken 20 years ago.

than 600× power. It is used in conjunction with high-quality Televue Barlow lenses to increase the focal length. At the rear of the telescope, the standard SCT focuser is "abandoned", and focusing is accomplished through a high-quality Crayford focuser manufactured by JMI, which is equipped with a digital readout of the focus position. This allows very fine and precise focusing.

2. Celestron 9.25-inch (23.5 cm) XLT Schmidt Cassegrain telescope: This telescope was acquired new in February 2004. The setup is very much the same as that on the larger scope, with a JMI motorized Crayford focuser being used to replace the standard one. This OTA can be used on either the large G11 or a smaller modified CG-5, which is used for travel purposes. This telescope has also provided many fine views and images.

However, these telescopes were both sold, being replaced, at the time of writing, with an Intes 10-inch F/6 Maksutov-Newtonian as the primary planetary telescope and a Meade 127ED Apochromat being employed for high-resolution imaging of the Sun.

Three different CCD cameras are used in all. I started with an SBIG ST-5c cooled CCD camera, but during late 2002 I made "the move" to using two webcams. These are the popular Philips TOUcam Pro webcam and, more

Figure 7.2. The author and his 28cm scope ready for action, with the imaging PC located close by. At the rear of the scope, the ATK-1HS is seen attached.

recently, an ATK-1HS black and white type. The webcams have allowed images of even greater resolution to be obtained with amateur telescopes (see Figure 7.1). Primarily, my telescopes are employed for long-term studies of Jupiter, Saturn and Mars, but I have also employed them for high-resolution imaging of the Sun (see Figure 7.3), Moon and binary stars.

Preparing to Acquire the Raw Images

Rather than describe the procedure used for cooled CCD cameras, I will instead detail the process I use for obtaining images with the webcams, which I have now used almost exclusively for the last 18 months. Today it remains a fact that the finest quality images of the planets are now produced with these devices, and currently no cooled CCD alternative to the webcam (that will produce as good results) is available. It is also of great note that color planetary images are much

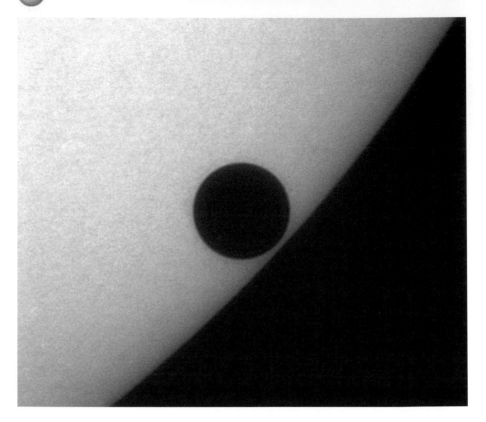

Figure 7.3. Today's webcams can also produce great results when imaging other objects, such as the Sun. This shot from June 8, 2004, shows Venus in transit across the solar disk. An 80mm Apochromat with Solar filter was used for this shot.

easier to obtain with webcams than the tricolor process of cooled CCDs; however the modern black and white webcams with sensitive CCDs can employ a wide range of filters.

After the telescope has been prepared (having reached ambient temperature, precisely collimated, etc), the observer is ready to begin imaging the target planet. I cannot emphasize enough the importance of critical collimation. A good way to think of the process is to ask the question: "would a pianist continue to use his/her piano without it being properly tuned?" Such is the case for getting the most out of any telescope – it must be properly maintained and collimated for it to produce the best possible results, when the seeing allows.

The collimation process is used when imaging involves the observations of a star at high power (typically at greater than 600×). The concentricity of the diffraction pattern in and out of focus is then observed. The pattern should appear completely concentric with a small bright dot at the center, surrounded by a series of bright and dark rings (see Figure 7.4). The pattern should close itself into

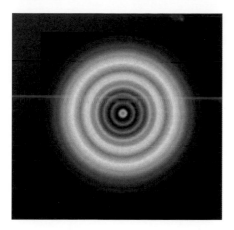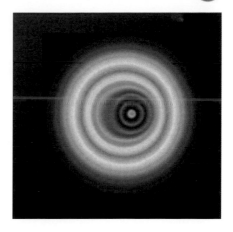

Figure 7.4. The appearance of a perfectly collimated (*left*) and mis-collimated telescope (*right*). Even a slight misalignment will result in a loss of contrast.

a symmetric Airy disk at focus and, in good seeing, with no obvious asymmetry. It should be noted at this stage that checking the collimation once a month, or once every few months, is completely unacceptable for high resolution work (especially if the telescope is transported any great distance by car). Also checking the collimation at low powers will not reveal misalignment – only high-power viewing will do that.

Before imaging can begin, the observer must choose an appropriate focal length to image at. In general, the Nyquist theorem dictates at least two pixels of the CCD must cover the theoretical resolving capability of the telescope. While this is true, the actual resolving capability of the telescope, on a planetary disk, is not the same as the Dawes or Rayleigh values quoted for limiting resolutions. Resolution of planetary detail is dependent on its contrast. For example, the tiny Encke division in Saturn's "A" ring spans just 0.05 arcseconds in angular width but has been imaged with apertures as small as 8 inches (20 cm), which is some 10 times smaller than the Dawes limit for this aperture. This is because the division represents a very dark line on a bright background, while the limit for resolving lower-contrast detail (such as small spots on Jupiter) will be nowhere near as fine as this level but still in excess of the Dawes value under good seeing.

Therefore, for telescopes in the 10–20 cm range, a sampling of around 0.50 to 0.25 arcseconds/pixel is high enough for typical conditions, while under excellent seeing I would recommend nearer 0.1 arcseconds/pixel with larger apertures (25–40 cm). Increasing the focal length is quite simple; I use high-quality Barlow lenses for the task, but eyepiece projection is also quite straightforward. For Jupiter I work at an image sampling between 0.19 and 0.13 arcseconds/pixel, while for Mars and Saturn nearer to 0.1 arcseconds/pixel. The formula you can use to work out the appropriate sampling is:

Pixel size of CCD (microns) / Focal length (mm) × 206 = arcseconds/pixel.

Shooting the Video

For shooting webcam videos using my TOUcams and ATK cameras, I use either the Philips Vrecord software supplied with the camera (see Figure 7.5) or the processing software IRIS. Both have straightforward control systems, and accessing the parameters is quick and easy.

First, the observer should click on the appropriate command to view a live streaming preview image from the camera and approximately focus the image. This will now be used to set the various parameters and later the fine focus of the image. Do not select frame rates higher than 10fps as this will result in very heavily compressed raw frames. I use either 5fps or 10fps, and these settings produce excellent results. The camera gain setting should be between ~20% and 70% for good results (adjust the slider bar in the camera software to choose the right level). Once these settings are defined, spend time carefully focusing the live video. If the seeing (and telescope!) is good you will easily note the point of precise focus. When imaging at lower altitudes (below 40°) the observer should insert an IR rejection filter, or the AVI frames will be smeared by atmospheric dispersion effects.

Since the planets rotate at different rates, there is a limited time in which to acquire all the frames for an image, which varies from planet to planet. This

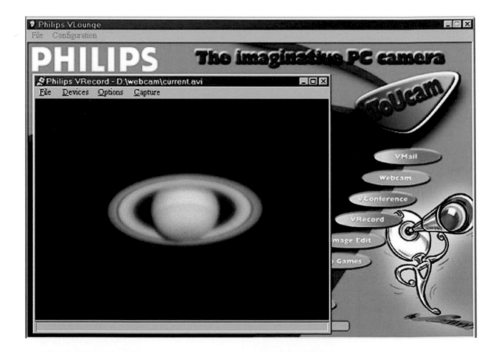

Figure 7.5. This is the screen visible ready for shooting AVIs using the Philip's Vrecord software. All the camera functions are controlled from here.

means around ~2 minutes for Jupiter and ~5 minutes for Saturn and Mars. For objects such as the Moon and Sun the imaging window can be much longer so this isn't really an issue. Venus also has a much longer imaging window.

Processing the Video

Two programs are available to the observer for the processing of webcam AVI files. Registax Version 2 (soon Version 3) and K3CCDTools are the primary programs I use for the processing of webcam images. Both are available free of charge for download.

While both programs offer many features, I would recommend Registax to the newcomer, as its user interface is easier to get to grips with and it offers a wide range of processing features in Version 2. All of my own images presented here were processed primarily with this program.

First, when the AVI has been loaded into Registax, you must sort through all the frames and select the best frame (least distorted by the seeing) as a reference. The program will then sort through all the frames and grade their quality. Before clicking to align the frames, it is important that the correct program settings are used, as indicated in Figure 7.6.

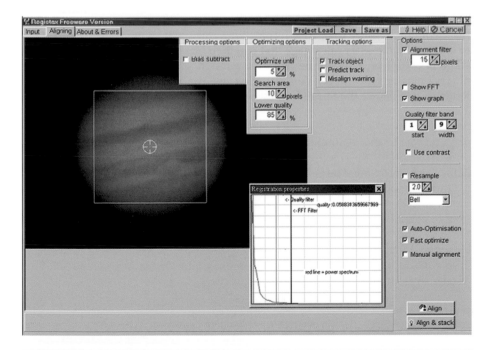

Figure 7.6. The two initial processing screens of Registax. It is essential that both the FFT filter (blue line) and the quality filter (green lines) are set to optimum values.

Figure 7.7. The final master raw image before and after enhancement using the wavelet sharpening routines in Registax.

Once the program has aligned all the frames, you are ready for it to average the raw frames into a final composite image. Typically anywhere between ~50 frames for the Moon to more than 1000 frames for Saturn will be used during this stage to create a final master raw image that can then be sharpened. When the stacking stage is complete, you will be left with a much smoother (though still rather blurry) master raw image. You can then use the wavelet sharpening tools in Registax to enhance the image further (see Figure 7.7). Caution must be exercised here, as it is very easy to get carried away and oversharpen the image (a very common occurrence among amateur astro-imagers). If the image doesn't respond well to sharpening, the data are not good enough and you must try again on a better night. No amount of post-processing will make up for mediocre raw data, so attention must be paid to previous stages!

Once processed, one must label the image with the appropriate data (such as date/time/telescope/seeing). Once completed, it is well worth submitting the images obtained to organizations that run planetary observing programs, such as the BAA and ALPO. Large amounts of data are collected from amateur work today, and with the resolutions now being obtained, it has become possible to follow the activity of planets such as Mars and Jupiter in tremendous detail, producing work valuable to professional researchers.

Practice Makes Perfect

Overall, one should not expect to achieve wonderful results on the first attempt – I can say this from experience. It will take patience and practice to finally start

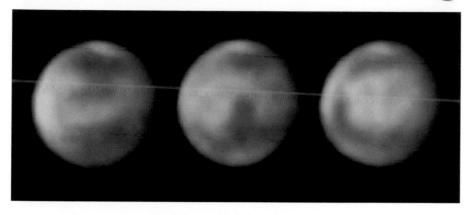

Figure 7.8. Mars showing clear surface detail, even though it subtended an angular diameter of just 4 arc seconds. With practice, capturing such detail can be very rewarding and useful.

achieving high-quality results. With patience, such challenging targets as imaging the surface features on Mars when only 5 arcseconds in diameter (see Figure 7.8) or detail on the Jovian moons become possible. The most rewarding part is, however, simply being able to follow the weather systems of other worlds with your own telescope and camera in the comfort of your own backyard. Today, large telescopes are not required to start getting useful results. Even high-quality telescopes as small as 100mm (4 inches) can produce decent images of the major planets and superb images of the Sun and Moon.

I hope this chapter will go some way to proving that producing detailed images of the planets is well within reach of today's amateur astronomers, even those on a limited budget. With technology so affordable, there has never been a better time to start studying the Earth's nearest neighbors.

CHAPTER EIGHT

High-Resolution CCD Imaging

Brian Lula

Using telescopes as small as 75mm aperture, amateur astronomers are now producing stunning images rivaling those of professional observatories of just a few years ago. The public has been fascinated by the dramatic and colorful high-resolution images taken by the Hubble Space Telescope, yet recent amateur astronomical images have been no less inspiring. Over the last few years the art of amateur imaging has made a quantum leap forward with film-format-sized CCDs, professional-quality large-aperture telescopes and sophisticated image-processing techniques all arriving on the scene. Just as important, amateur astronomers have learned how to master them.

This chapter discusses high-resolution deep-sky CCD imaging, which is one of the most technically challenging aspects of astrophotography. Fundamental to producing stunning color CCD images is fully understanding what contributes to acquiring the best possible images. The bottom line is that the better the raw image, the better the end result. Like any skill or craft it becomes easier as you understand the requirements, prepare your equipment, practice and then experiment with the techniques. The inspiration will be the many images acquired as you advance along the learning curve. These images will reveal the heavens as you've never seen them before.

To become proficient at this facet of CCD imaging we will address the following questions:

a. What is high resolution?
b. How do I prepare for high resolution?
c. Why is "seeing control" important?
d. What demands does it place on equipment?
e. What demands does it place on acquisition?

Figure 8.1. Bubble Nebula, NGC7635. (Ha)(Ha/R,G,B) image of 180:30:30:30 minutes total exposures, respectively (3 minute individual exposures for the RGB, 5-minutes for Ha) using a 3nm Ha Custom Scientific Ha filter for the luminance frames and Custom Scientific RGB filters. The 180-minute Ha image was combined 50/50 with the R frame in Maxim DL to provide an enhanced R frame using the discrete Ha emission. Image calibration and color combination using MAXIM DL (RC Console for Sigma Combine and pixel cleanup within Maxim DL), AIP for Lucy Richardson deconvolution on the luminance, image registration using Registar, Ron Wodaski's gradient removal, luminance layering in Photoshop for final color and star shaping processing. Equipment: RCOS 20-inch f/8 RC and Finger Lakes IMG6303E CCD camera with all images acquired in 2 × 2 bin mode for an image scale of .92 arcsecond/pixel in 3 arcsecond seeing and magnitude 4.9 suburban/rural skies.

What Is High Resolution?

The term high resolution has changed over the last 10 years. Generally it means working at image scales where seeing dominates the ability to resolve fine detail in an image. It can range from 1.0 arcsecond/pixel image scales under 3 arcsecond seeing conditions to 0.5 arcsecond per pixel for 1.5 arcsecond or better conditions. I have been imaging with CCDs since 1995, starting with the construction and operation of a highly modified Cookbook 245 CCD camera. At that time it was stated that 2 arcsecond/pixel was the optimum image scale for deep-sky imaging under average seeing conditions.

By the late 1990s, images taken by Rob Gendler and Bill McLaughlin demonstrated noticeable resolution improvements by working at 1 arcsecond/pixel or

Figure 8.2. Galaxy NGC4244. LRGB image of 80:50:50:70 minutes total exposures, respectively (5-minute individual exposures) using a non-IR-blocked luminance image and Custom Scientific RGB filters. Image calibration and color combination using MAXIM DL (RC Console for Sigma Combine and pixel cleanup within Maxim DL), image registration using Registar, Ron Wodaski's gradient removal, AIP for Lucy Richardson deconvolution on the luminance, luminance layering in Photoshop for final color and star shaping processing. Equipment: RCOS 20-inch f/8 RC and Finger Lakes IMG6303E CCD camera with all images acquired in 2 × 2 bin mode for an image scale of .92 arcsecond/pixel in 2.3 arcsecond seeing and magnitude 4.9 suburban/rural skies.

less image scales. This coincided with the availability of smaller pixel CCD arrays, such as the Kodak KAF1600 and KAF0400 with their 9u pixels followed by the KAF3200 with its 7.4μ pixels. The tradeoff was longer exposure times but the results proved that significant resolution improvement could be obtained. At the same time, new image-processing techniques, especially in the area of deconvolution, allowed the extraction of high-resolution information that was hidden by the seeing blur in the image frames.

These developments resulted in deep-sky color images taken by amateurs that revealed previously unrecorded detail and color information. Currently the challenge for advanced imagers is finding reference images for comparison of color balance or that some new feature in the image is actually real. For example, in early 2004, I imaged M106, a very large galaxy with faint outer arms, using my 20-inch RCOS scope and Finger Lakes IMG6303E CCD camera (Figure 8.3). I took a long H-alpha exposure to best capture some of the discrete HII regions in the arms and followed it with a conventional clear luminance series and RGB color set. During the final color combination steps I spotted two noticeably red colored

streaks originating from the core of the galaxy. I checked a number of other personal imager Web sites and couldn't find any record of them. With further research, I found a weak monochrome image on a professional observatory Web site that confirmed the feature. Sure enough they were galactic jets, something unexpected in M106, and finding them was a first for an amateur.

How Do I Prepare for High Resolution?

Like real estate, the old adage of "location, location, location" rings true for obtaining the highest resolution CCD images. Some areas are simply better than others. Mountaintops in the desert areas of the Southwest United States, Southern Africa and South America have the highest number of nights of very dark, sub-2-arcsecond seeing. Most of us have to accept what lies overhead, whether it is turbulence from the Rockies or the air-mixing effects of powerful jet streams. These atmospheric phenomena are part of the macro-seeing environment, and they are totally outside our control. Surprisingly, though, we can significantly control the "local" or micro seeing environment for imaging, but we understand that everything conspires against us working at sub-arcsecond/pixel image scales!

Here are a few lessons based on my experience:

Rooftops: Homes heat up during the day and take a long time to cool down, especially with roofs made of ceramic tile or asphalt shingles and where the structure is brick or stucco. Do not select a target downwind of your home unless you have allowed adequate time for cool-down. My observatory is about 70 feet from my house, and it is not practical to image until about 3 hours after sundown if my target is downwind of the house.

Chimneys: Streams of chimney exhaust travel surprisingly long distances. This meandering stream can cause the loss of guide-stars and create large variations in your star sizes if it crosses over the area imaged by your telescope. A few very prominent imagers have confessed to turning off their furnaces at night, even in the dead of winter, while imaging. Tip: for family unity put up a reminder to turn it back on when you've finished!

Figure 8.3. Galaxy M106. L(Ha/R)GB image of 150:50:50:70 minutes total exposures, respectively (3-minute individual exposures) using a non-IR-blocked luminance image and Custom Scientific RGB filters. A separate 90-minute Ha image (5-min. individual exposures) was combined 50/50 with the R frame to better highlight the galactic jets. Image calibration and color combination using MAXIM DL (RC Console for sigma combine and pixel cleanup within Maxim DL), image registration using Registar, Ron Wodaski's gradient removal, AIP for Lucy Richardson deconvolution on the luminance, luminance layering in Photoshop for final color and star shaping processing. Equipment: RCOS 20-inch f/8 RC and Finger Lakes IMG6303E CCD camera with all images acquired in 2 × 2 bin mode for an image scale of .92 arcsecond/pixel in 3 arcsecond seeing and magnitude 4.9 suburban/rural skies.

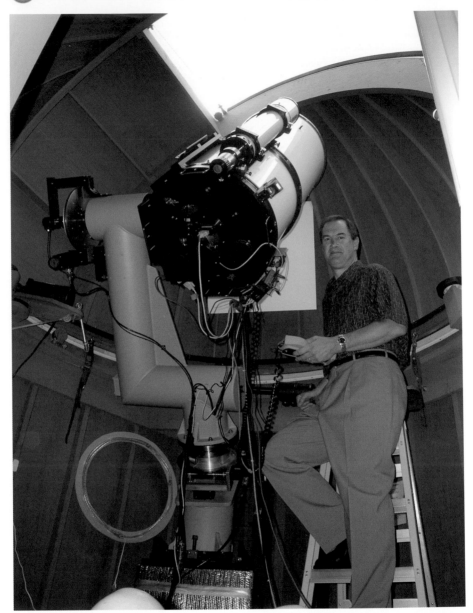

Figure 8.4. Brian Lula with his 20-inch Ritchey-Chretien (RC) telescope and home-built heavy-duty equatorial fork mount. This telescope replaced a personally built 20-inch F/5 Newtonian astrograph used to take some of the deep-sky images in this chapter. Note the silver wrap on the telescope pier. The pier produced considerable thermals during the night, seriously affecting local seeing. An inch-thick layer of rigid Styrofoam glued all around the pier and further wrapped by aluminized polyethylene bubble wrap improved seeing conditions considerably.

Asphalt parking lots and driveways: These are also strong heat absorbers that will continue heating up air long after sunset. Avoid imaging downstream or on top of these areas until they have adequately cooled down. Large grassy areas are optimum.

Home observatories: Modern professional observatories are especially designed to alleviate local seeing effects produced by the telescope enclosure. Not surprisingly, the same applies to our personal observatories for high-resolution work. Open up domes or roll off roof observatories well before sundown to give them plenty of time to cool down. Even better is to pull air into your observatory with fans. Avoid cinder block or brick walls and concrete floors: They will absorb significant amounts of heat during the day and expel it for long periods during the night. Insulate large concrete or steel piers that support the telescope. I use a rigid styrofoam insulation board to insulate the concrete pier and further wrap it with aluminized "bubble wrap" to protect it (see Figure 8.4).

We can spend considerable effort and money managing the thermal behaviors of our telescopes with fan cooling and zero expansion optical/structural materials. However, overall imaging performance can be as much affected by local seeing phenomena as by our scope performance.

Why Is "Seeing Control" So Important?

With the goal of extracting the highest resolution possible in our CCD images, seeing plays a much bigger part than is at first realized. Longer exposures do not mean you can detect an object if the seeing is poor. At the 2004 Riverside Telescope Makers Expo, Jim McGaha, an advanced CCD imager specializing in Near Earth Object (NEO) detection, gave a talk centered on detecting objects fainter than 20th magnitude. Using information presented in his talk, Table 8.1 reveals how a much fainter star can be detected with any given aperture of telescope as the seeing improves.

Table 8.1. Effect of Seeing on Detection Magnitude

Seeing in arcsecond (FWHM)	Improvement in Detection Magnitude (using 4 arcsecond seeing as baseline)	Aperture Required to Achieve Magnitude Increase (using a 6″ aperture scope as baseline) (inches)
4.0	0	6
3.5	0.7	9
3.0	1.4	12
2.5	2.1	16
2.0	2.8	20
1.5	3.5	30

Figure 8.5. Iris Nebula, NGC7023. LRGB image of 120:30:30:45 minutes total exposures, respectively (3-minute individual exposures) using a non-IR-blocked luminance image and Optec RGB filters. Image calibration and color combination using MAXIM DL, image registration using MAXIM DL, AIP for Lucy Richardson deconvolution on the luminance, luminance layered in Photoshop for final color and star shaping processing. Equipment: Homemade 20-inch f/5 Newtonian astrograph and Finger Lakes IMG6303E CCD camera with all images acquired in 2 × 2 bin mode for an image scale of 1.48 arcsecond/pixel in 3.2 arcsecond seeing and magnitude 4.9 suburban/rural skies.

What this means is that telescopes in the 6- to 8-inch range, when used optimally in good seeing conditions, will perform as well for deep-sky imaging as much larger scopes under average to poor seeing (see Figures 8.5 and 8.6). Seeing levels the playing field for most of us. This is very encouraging to imagers with telescopes in the 4- to 8-inch range. They can capture truly stunning images when the macro seeing conditions are favorable, providing they have taken the proper steps to manage their local "micro" seeing environments. The bottom line is that you can economically improve imaging performance by controlling your local seeing environment without spending large sums on more aperture!

It is common practice now for many imagers living under generally turbulent skies to only take luminance images during nights of good seeing and then take their color data on an average seeing night. This is due to the fact that the luminance image provides most of the resolution data in the final color composition.

Figure 8.6. Iris Nebula, NGC7023. Comparison image using 6-inch scope This remarkable image by Cord Scholz of NGC 7023 compared to the one taken by the author with his 20-inch in similar seeing conditions shows what is achievable with modest aperture scopes used skillfully for deep-sky color imaging. LRGB image of 210:30:30:50 minutes total exposures, respectively, with 5-minute individual exposures in L and 10-minute in RGB. Image calibration and color combination using MAXIM DL, image registration using MAXIM DL, RC Console for sigma combine and pixel cleanup, AIP for Lucy Richardson deconvolution on the luminance, LAB combined in Photoshop for final color and star processing. Equipment: 6-inch f/6 Intes Maksutov with SBIG ST-10XME and CFW8 filter wheel in 3 arcsecond average seeing.

The Demands on Equipment

As imagers advance their techniques, they become more critical of star shapes in their images. With sub-arcsecond/pixel image scales they find that keeping them tight and round into the extreme corners places a significant demand on their entire equipment setup. With film-format-sized CCD arrays, many scopes need to be modified or even replaced to obtain the quality of images one strives for. Most telescopes deliver fields adequate for the midsized series of CCD cameras, such as the SBIG ST-8 and -10 series. However, before taking the plunge for bigger CCDs, consider if the performance is good enough or can be made good enough for wider field work.

As we start the discussion on equipment requirements we should first put the tolerances of high-resolution imaging into context. Let us use the KAF3200 CCD array as an example. The individual pixels in this popular CCD are 6.8 microns square. The human hair has a diameter of 75 microns, so that means it takes 12 pixels to span the tiny width of a hair. It is immediately obvious how little mechanical movement or optical aberration would smear a faint star image spanning just a few pixels. These small errors are invisible, which makes it so difficult to track down the root problem.

Here is how I prioritize the importance of individual components in the imaging system:

Priority 1: Mount Quality. More deep-sky high-resolution images are ruined by poor mount performance than optical quality. Even on a tight budget, do not cut corners on the mount. Good mounts are harder to find than good optics, are usually more expensive to buy and are even harder to make (see Figures 8.7 and 8.8). The necessary tools and machinery to make them are not commonly available. For ATMs, making good optics requires surprisingly unsophisticated equipment. However, a mount requires precision machine tools (and machining skills)

Figure 8.7. Brian Lula and portable imaging setup. Reinforcing Brian's opinion about mount rigidity for CCD imaging is his design of a portable German equatorial mount for scopes up to 16-inch diameter with long focal lengths. It is shown here at the Stellafane Amateur Telescope Making Conference.

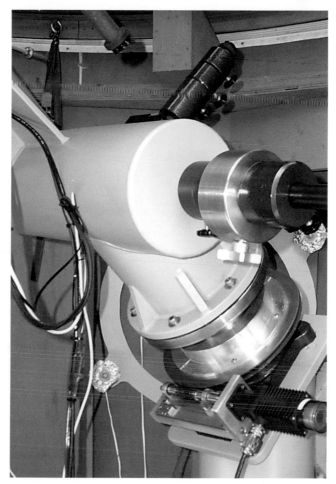

Figure 8.8. A mechanical engineer by profession, it shows in Brian's home-built heavy-duty equatorial fork mount. Home-building a precision-tracking mount is an onerous task. Note the silver wrap insulation on the telescope pier.

to obtain the necessary tolerances for it to track smoothly and accurately at the subpixel guiding accuracy needed for high-resolution CCD imaging. I have built a number of mounts for imaging with scopes up to 20 inches in aperture and 160-inch focal lengths. I learned the hard way how difficult it is to make a strong and accurate mount and endured many reworks to get it right. A "good" mount will give you a high yield of usable high-resolution images but it will also be a treat for visual observing, especially at high power.

Priority 2: Optical Tube Assembly (OTA). Many commercial OTAs have a high enough optical quality to produce excellent images but are surprisingly poor mechanically for CCD imaging. Many advances have been made, however, over the last couple of years, by progressive telescope and component manufacturers. They have added features such as:

- stronger focusers with zero backlash motion and image shift,
- primary mirror lock-down mechanics for Schmidt-Cassegrains,
- stiffer tube assemblies with carbon composite tube or truss designs,

- active cooling systems for more rapid cool-down and better thermal control,
- zero expansion optical materials to minimize mirror distortion during cool down,
- zero expansion structural materials, such as carbon composites or invar, to minimize focus shift and
- active focusers to compensate for image shift due to temperature changes.

A growing problem, as larger CCDs become more widely used, will be the quality of the image at the focal plane for a specific optical design and the aberrations that arise from poor alignment. High-resolution imaging requires longer focal lengths, which tend to point to reflecting telescopes. Collimation and aberrations are issues to manage with these designs. Even the highly regarded f/8 Ritchey-Chretien telescope cannot provide diffraction-limited performance over the large CCD arrays now used by a number of advanced amateur imagers. Field correctors and even new OTA designs will be needed to address this problem.

Figure 8.9. Cave Nebula, Sharpless 2-155. LRGB image of 120:30:30:30 minutes total exposures, respectively (3-minute individual exposures due to the presence of bright stars) using a non-IR-blocked luminance image and Custom Scientific RGB filters. Image calibration and color combination using MAXIM DL (RC Console for Sigma Combine and pixel cleanup within Maxim DL), image registration using Registar, Ron Wodaski's gradient removal, AIP for Lucy Richardson deconvolution on the luminance, luminance layering in Photoshop for final color and star shaping processing. Equipment: RCOS 20-inch f/8 RC and Finger Lakes IMG6303E CCD camera with all images acquired in 2 × 2 bin mode for an image scale of .92 arcsecond/pixel in 2.8 arcsecond seeing and magnitude 4.9 suburban/rural skies.

A final note on the subject of OTAs. Really understand and practice collimation if you are using a reflecting or catadioptric design (i.e., Schmidt-Cassegrain). Just as you can increase the effective size of your telescope by improving your local seeing, so too does achieving good collimation. Many excellent Web resources are around to help you get the best collimation.

Priority 3: Focuser. Obviously you cannot image without a camera but without a responsive and rock-steady focuser you won't want to image anyway. Focusing at high resolution requires patience, care and practice. In the micron realm of focusing, the whole OTA structure is constantly on the move due to temperature changes and structural flexure, so accurate focus control is paramount. Many robust focusers have been built specifically for CCD imaging with the Crayford design generally the most popular. Even this design has had to go through recent improvements to be strong enough to support heavy filter wheels and CCD cameras, without introducing tilt errors. Excellent motorization is available with products such as Robofocus, which allows focusers to be remotely controlled. These can even cater for nonparfocal filters and focus position changes of the optical tube assembly caused by temperature variations.

Figure 8.10. Galaxy M101. LRGB image of 90:30:30:45 minutes total exposures, respectively (3-minute individual exposures) using a non-IR-blocked luminance image and Optec RGB filters. Image calibration and color combination using MAXIM DL, image registration using MAXIM DL, AIP for Lucy Richardson deconvolution on the luminance, luminance layered in Photoshop for final color and star shaping processing. Equipment: Homemade 20-inch f/5 Newtonian astrograph and Finger Lakes IMG6303E CCD camera with all images acquired in 2 × 2 bin mode for an image scale of 1.48 arcsecond/pixel in 3.4 arcsecond seeing and magnitude 4.9 suburban/rural skies.

Recently a freeware product called FocusMax was introduced to control a number of industry-standard focusers and CCD cameras, which automates the entire focusing routine. It works extremely well and is amazing to watch as it optimizes your focus automatically!

Priority 4: The CCD Camera. Last but not least is the imaging camera itself. If you take care of the preceding priorities, almost any cooled imaging camera will provide first-rate deep-sky images. No matter the choice of manufacturer, important considerations are QE (light sensitivity), low noise electronics, adequate cooling power and stable cooling characteristics, stiff mechanical assembly, high reliability, software compatibility and service support. The price of cameras varies dramatically with CCD sensor size, ranging from less than $2,000 for a TE cooled Kodak KAF400 sensor based camera to more than $20,000 for a high-end

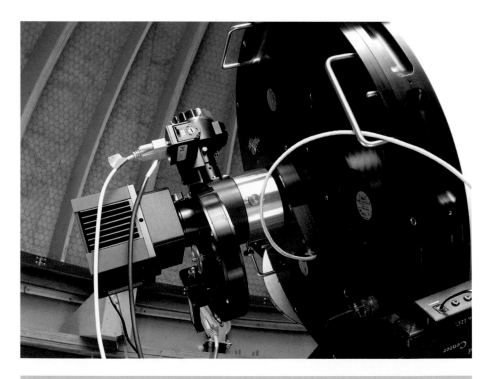

Figure 8.11. The imaging train. This photograph details the imaging train of Brian's 20-inch RC telescope system. Starting from the telescope back plate is a threaded instrument adapter to space the focal plane at the right distance from the back plate followed by an instrument rotator (the red component) to assist in framing a large astronomical object or finding a suitable guide star. Behind the rotator is a personally built 12-position motorized filter wheel (2–6 position wheels) with an off axis guide port and autoguider to greatly help in maintaining guide star lock during filter changes and simplify finding suitable guide stars. The autoguider has its own remote focus to maintain sharp guide star focus for nonparfocal filters or straight through imaging. The camera is a Finger Lakes IMG6303E (2K × 3K × 9μ pixels).

camera with a Kodak KAF 6303E CCD array (see Figure 8.11). The reality is that most astronomical objects are quite small, so smaller CCD arrays are more than adequate for most high-resolution imaging.

The Demands on Image Acquisition

Now that we have optimized our setup we enter the most enjoyable part: taking the images! Over the last 5 years one trend has emerged. Good images require much longer exposures than anyone at first thought necessary for CCD imaging. In the "early" days, which are only within the last 10 years, most image exposures were measured in tens of seconds and taken in groups (sequences) of maybe 10. Today, to obtain enough data to produce publishable-quality images, total exposure times can exceed 6 hours!

A topic of much discussion is the individual exposure time necessary to obtain good data, depending on whether you image from a suburban or dark sky. In general, the longer the exposure the better your image will be. However, in practice exposure times vary considerably, depending on the brightness of the target, the seeing, the size of bloomed stars, how much nighttime you have, the sky background and your mount's performance. Most high-resolution deep-sky color images taken today have minimum total LRGB exposure times of 60:20:20:25 minutes, respectively, with 2- to 3-minute individual exposure times (see Figure 8.12). Maximums can go up to staggering exposures of 240:60:60:80 minutes, respectively, with 10-minute individual exposure times.

Two hardware features greatly assist in acquisition of "keeper" images in high-resolution work. One is the off-axis guider, which is a separate CCD camera that "looks" down the same optical path as the main imaging camera and tracks the relative motion errors of both the optical system and mount. This is much better than using a guide scope, which, due to differential flexure between the guide scope and main imaging scope, makes it almost impossible to obtain consistently round star images with long exposures. SBIG manufactures CCD cameras with a guider chip built in to the main camera body, which overcomes the differential flexure problem. This has many merits but, because the guide camera images through the color filters, light to the guiding chip is much reduced. This is especially a problem when narrowband filters are in use. At subarcsecond/pixel scales it is much harder to find suitable guide stars; therefore, many imagers are beginning to use a separate off-axis guide camera *in front* of the filter wheel. The most important benefit is that the guide star is unattenuated no matter what filter is in use. In addition, there is a bigger area to "sweep" for a guide star when using a free-rotating device. The down side is the need for more back focal length to insert the off-axis pick-off tube. Maintaining focus at the autoguider with nonparfocal filters can also be a problem, but some scopes even have a remote autoguider focus to remedy this.

Figure 8.12. Emission Nebula, NGC6820. (Ha)(Ha/R,G,B) image of 96:30:30:30 minutes total exposures, respectively (3-minute individual exposures on the RGB, 5-minute for Ha) using a 3nm Ha Custom Scientific Ha filter for the luminance frames and Custom Scientific RGB filters. The 96-minute Ha image was combined 50/50 with the R frame in Maxim DL. Image calibration and color combination using MAXIM DL (RC Console for Sigma Combine and pixel cleanup within Maxim DL), image registration using Registar, AIP for Lucy Richardson deconvolution on the luminance, luminance layering in Photoshop for final color and star shaping processing. Equipment: RCOS 20-inch f/8 RC and Finger Lakes IMG6303E CCD camera with all images acquired in 2 × 2 bin mode for an image scale of .92 arcsecond/pixel in 3.8 arcsecond seeing and magnitude 4.9 suburban/rural skies.

Another useful tool is a high-speed tip/tilt mirror ahead of the main imaging camera to eliminate the short time scale disturbances to the image. These can be caused by seeing, wind buffet or telescope drive anomalies. This device counteracts object wander and produces a stationary image on the CCD. These devices do not discriminate between seeing effects, mechanical bumps or hiccups in the drive so the net result is much tighter star shapes. SBIG manufactures a device called the AO7, which does this job, and it is an important tool for advanced high-resolution imaging.

Summary

If you optimize the areas discussed you will be able to produce images that, once calibrated and combined, will provide the basis for producing those really top-notch high-resolution color CCD images we all aspire to. The beginning imager is wise to start with low-resolution wide-field work because it is considerably less challenging from a mechanical standpoint and it offers a less frustrating path to perfection than the deep-sky high-resolution imaging described here. But if you're up for the challenge, go for it!

CHAPTER NINE

Out-Smarting Light Pollution

David Ratledge

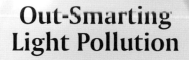

Introduction

My location in light-polluted Lancashire and my interests in the deep, deep sky would appear to be mutually exclusive. However, advances in techniques as much as advances in technology have meant that we can successfully image all manner of interesting objects (see Figure 9.1). We cannot expect to rival the best of images from dark sites but by acting smarter in subject choice, imaging

Figure 9.1. The power of CCDs. With just amateur equipment we are able to reach some of the most remote objects known – high-redshift (z > 4) quasars. Their light has taken around 12 billion years to reach Earth! FLI Maxcam and 16-inch Newtonian.

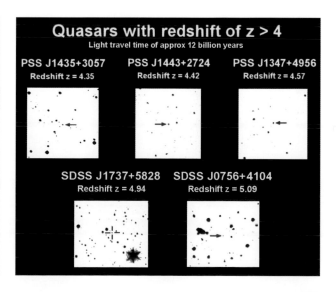

Quasars with redshift of z > 4
Light travel time of approx 12 billion years

PSS J1435+3057
Redshift z = 4.35

PSS J1443+2724
Redshift z = 4.42

PSS J1347+4956
Redshift z = 4.57

SDSS J1737+5828
Redshift z = 4.94

SDSS J0756+4104
Redshift z = 5.09

techniques and processing methods, we can fight back. Smarter subject choice means searching out suitable targets – there are plenty, believe me. Smarter imaging techniques means optimizing filters and exposures for the limits of our site. Smarter processing includes using state-of-the-art noise-reduction filters – but more on that later.

Equipment

At first glance the equipment needed to successfully carry out CCD imaging seems incredibly complex with all manner of electronic gizmos and wires everywhere. While it is possible to image by setting up the equipment every night, life is certainly simpler if the equipment, or the majority of it, can be permanently set up and wired together in an observatory. For me, an observatory provides another vital function – it helps to mask neighborhood lighting. While general light pollution has to be endured, backyard lights shining directly into the telescope tube should be avoided. A classical dome helps with this, as just the slit is open (see Figure 9.3), and I can even reduce the opening further by using strategically placed plastic sheets. They don't have these problems on Mount Palomar!

Figure 9.2. David Ratledge at his observatory.

Figure 9.3. 16-inch telescope inside the observatory. A classical dome is ideal for blocking local light pollution. The lower computer controls the telescope while the upper laptop operates the CCD camera.

With my interest in the deep sky, a big fast Newtonian telescope was the obvious choice. It is also the easiest to home-build. It was a joint effort by a team of three: Brian Webber made the optics, Gerald Bramall made the tube assembly and I made the mount and electronics. A 16-inch (40cm) was chosen as a good practical size with a fast focal ratio of f/4.7. This translates to a nominal focal length of 75 inches (1900mm), which was a good match for the CCD I was likely to use (and afford). At this focal ratio, imaging would be quick and coma would not be too objectionable over the central 10mm. For bigger chips I would need a coma corrector, but that's a problem for the future. The tube assembly was designed with light pollution in mind and is of an aluminum-framed construction. It features a baffled design – the opening in each baffle increases away from the mirror matching the telescope's field of view – the idea being to trap internal reflections that would reduce contrast. The tube frame was finally covered in thin plastic to further reduce light entry. My pride and joy on the telescope is an 80mm-diameter electric Crayford focuser (see Figure 9.4). This is a dream to use with absolutely no backlash, and focusing to 1/1000 inch is straightforward!

With the light pollution that blights my site, virtually every object I image is invisible even in the 16-inch telescope. A flip mirror finder is therefore of no use in locating objects. The telescope is capable of imaging just about everything but I couldn't locate anything! I had therefore to build a computer GOTO mount to overcome this. This was based around Comsoft's PC-TCS software running on an old DOS PC. Big 48-volt stepper motors are employed to drive and effortlessly slew the telescope around. It was a relatively simple task to wire up all the components and make a hand control box. This has been a huge success and I can now acquire any object, visible or not. Also mounted on top of the main tube is a Meade 6-inch (150mm) Schmidt Newtonian, which is used for wider fields of view.

I have been somewhat less successful with CCD cameras. However, my original HiSIS22 camera based on a Kodak KAF400 chip (768 × 512) is still going strong after nine years. With its 9-micron pixels it is still my first choice for use on the 6-inch telescope for objects like comets and open clusters. The main telescope was designed with a Kodak KAF261 in mind. This 10mm square chip has 20-micron square pixels (512 × 512) and is a good match to the 75-inch (1900mm) focal length. After two troublesome cameras (from different manufacturers) I am now having better luck with a Finger Lakes Instruments MAXCAM camera (see Figure 9.4). This is a relatively compact camera, and results have been good. My camera was one of the first batch with a USB connection, which downloaded images in a leisurely 6 seconds. It was subsequently upgraded to a high-speed USB interface with downloads of less than 1 second! Focusing is now like having a video camera.

My final piece of kit is a filter holder. Initially I used a homemade filter system with twin wheels mounted close together. This permitted more combinations of filters than is possible with a single wheel. However, when two filters are used together, reflections between them can be a problem, especially if a bright star is present in the image. I replaced it with an Astronomik manual filter drawer and I now just use a single filter. To combat light pollution I generally image with a simple red filter in place. This not only passes the red part of the spectrum but

Figure 9.4. FLI Maxcam CCD camera. Note the large Crayford focuser. Pencil marks on the side of the focuser might look crude but they indicate the approximate focus position.

the (near) infrared as well, where CCD chips are still very sensitive. If the object is blue, then I use a Lumicon Deep-sky filter which, in addition to the red, also passes at the blue end of the spectrum.

Taking Procedure

What are those smarter imaging techniques I referred to earlier? The method I have evolved over the years involves taking short exposures – and lots of them! This has many advantages, which more than outweigh the disadvantages. The first is in combating light pollution. Long exposures would simply swamp the CCD at my location. Second, short exposures mean no guiding or even the hassle of having to set up an auto guider. Third, should anything go wrong, such as an airplane crossing the field of view or the telescope being bumped, then only one short exposure is lost. Fourth is bleeding, those unsightly vertical streaks from bright stars, which is much reduced by short exposures. The alternative, anti-blooming gated (ABG) chips, come with a penalty of reduced quantum efficiency of up to 30%. This is too high a price to pay. The disadvantage is, of course, that every time an image is read out it suffers degradation (noise), and the more images we have the more readout noise we have.

What is a short exposure? That really means an exposure that is long enough to record a meaningful signal. In my case, with fast f/4.7 optics and big 20-micron square pixels it means 15 to 60 seconds for most objects. I have standardized on 30 seconds as a good compromise. The number I take varies depending on the type of object – 33 exposures are enough for star clusters, 66 for galaxies and 99 for anything really faint. These may seem like big numbers, and from a dark site we could get away with a lot fewer but, from a light-polluted site, the only way to get a decent signal-to-noise ratio is to increase the number of exposures.

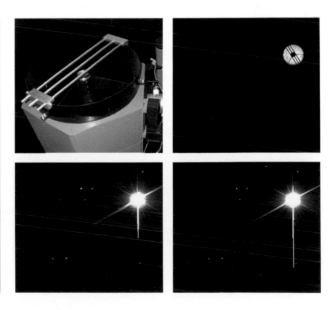

Figure 9.5.
Diffraction focusing using triple bars in front of the telescope. Image a bright star and adjust the focus until a single slender long spike is visible. Top right – a long way from focus; bottom left – getting close to focus; bottom right – at focus.

Before beginning an imaging session I prepare finder charts using a sky-charting program (SkyMap). Not only is this essential for confirming I am aiming at the right point, but it is very useful for framing. The program allows the positioning, on the finder chart, of the precise CCD field of view – in my case an 18-arcminute square. In that way I might be able to position the field of view to include other objects in the vicinity, such as faint background galaxies.

At the telescope, the first job after booting everything up is to align the telescope on a bright star to initialize the GOTO. This bright star is also used for focusing. I have always used diffraction focusing and found it simple and reliable. To do this I place three parallel bars across the front of the tube – the central bar I align exactly with one of the diagonal mirror supports. Then, with no binning (i.e., 1 × 1) and the visualization set to "MAX," I adjust the electric focuser until a single bright spike is visible (see Figure 9.5). It's as simple as that. The method is absolute. When you see a single slender long spike you know it's at focus. There's no need to try to see if it's better by pushing the focus button once more. The tricky bit is to remember to take the bars off!

The software I use for controlling the camera is MAXIM/CCD (see Figure 9.6). Its operation is very logical. It has a "sequence" mode whereby a generic name for the file is entered and the instruction is given for automatically taking a number of exposures one after the other. As mentioned, I set this to 33. I can then go indoors, and the software will take the exposures and save them, numbering each incrementally. This number of exposures is a practical limit for my setup – after this many the dome needs turning and, if there is any drift, I recenter the object before setting it off on another 33.

I take calibration frames the following evening, providing it is not raining! This does not waste imaging time, which is extremely valuable in Lancashire where clear nights are few and far between. I take 33 of each, i.e., bias, dark and flat. This is, of course, only possible because I can leave the camera on the telescope in the observatory untouched in the exact focus and orientation I used the night before. The bias (zero exposure) and dark (30-second exposure), both with the telescope capped, are straightforward. The flat I take of the evening sky – either clear or

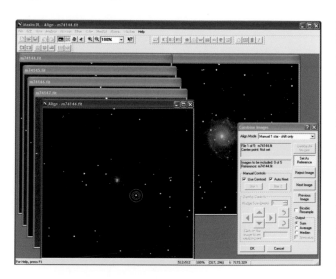

Figure 9.6. MAXIM being used to combine multiple exposures. A single star is selected for aligning all the images. If the images had been taken on different nights, with the camera in a slightly different orientation, then two-star alignment would have been used.

cloudy – and time it to when it is just dark enough for a 1- to 2-second exposure to register about 70% maximum brightness. If clear, a few stars are usually recorded, but by taking a median these will be removed.

Image Processing

Using a laptop computer in my observatory means that for processing I have to get all the files, i.e., images and calibration frames, downloaded into my desktop computer. For this I use an ethernet network, which transfers the files in just a few minutes. When they have been backed up to CDROM, they are deleted from the laptop.

For many years I used QMIPS32 for calibrating and processing images but I have switched to a combination of the freeware IRIS and MAXIM. The latter software is not cheap but makes calibration quick and easy, especially for my methodology of taking tens of exposures of each single object. If several objects were taken during an evening, it can create master bias and dark and flat frames, further speeding up processing. Each individual frame is calibrated separately before being "aligned." I do the latter using the single-star option, i.e., every frame is loaded in turn and simply by pointing at the same star on each, MAXIM

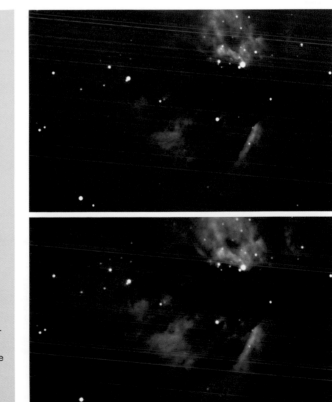

Figure 9.7. A rather extreme case of noise reduction. The Flame Nebula, taken with no-light-pollution-blocking filter on a C8 telescope (*top*), rescued with a wavelet filter in IRIS Software (*bottom*).

works out the offset and then adds all the frames together. The dynamic range that results considerably exceeds the traditional 64,000 limit, but MAXIM can handle this by saving files in 32-bit FIT format. Before saving the summed frame I remove the approximate background level, erring a bit on the low side so I don't inadvertently remove faint detail. I can then save the image as the calibrated frame.

However, actual image processing cannot yet begin. Because of the severe light pollution there will inevitably be a background gradient across the image. This varies according to where my telescope was pointing and which neighbor had his outdoor lights on! MAXIM has "flatten background" and "remove gradient" commands but I have found these unable to cope with the complexity of my light-polluted images. The only program able to rescue the image from the fog that I have found is IRIS. This has a command to synthesize a background manually (it has an auto command but manual is usually better). To do this you simply "point" at the background, carefully avoiding stars and the object of interest. Two hundred or three hundred points are usually enough. A polynomial best-fit back-

Figure 9.8.
NGC2403. Color image using the LRGB method with just a red and blue image. The green was synthesized by combining the R and B and the R used as the L. Top left – red image; top right – blue image; bottom-final LRGB image.

ground can be synthesized and saved. This is then simply subtracted from the calibrated image.

At last, image processing can begin. I usually try two options and see which works best. The first is Digital Development (DD) using MAXIM. This tends to be good with faint tenuous objects such as the outer reaches of galaxies, but it also holds detail right to a galaxy's core. If not carefully carried out it can produce dark rings around bright stars. No further processing is usually required with this option. The second option is Richardson-Lucy (RL) deconvolution using IRIS. This sharpens without producing dark rings round stars. It is an iterative process and restores the image based on a selected star within it. I generally only use 5 to

Figure 9.9. Flaming Star Nebula IC405. This color image is a combination of a full-resolution CCD mono image and low-resolution color information from an old 35mm slide.

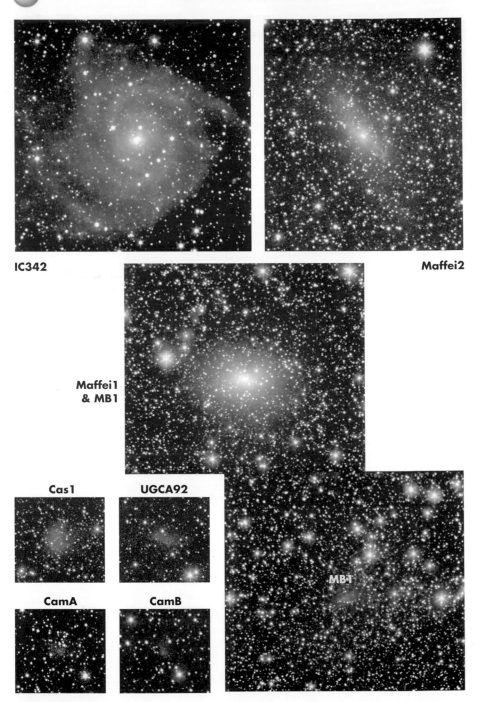

Figure 9.10. IC342/Maffei Galaxy Group montage (1). This group has some of the most obscured galaxies known and it is only relatively recently that several members have been discovered. Yet our backyard scopes, under light-polluted skies, can successfully image them in the infrared.

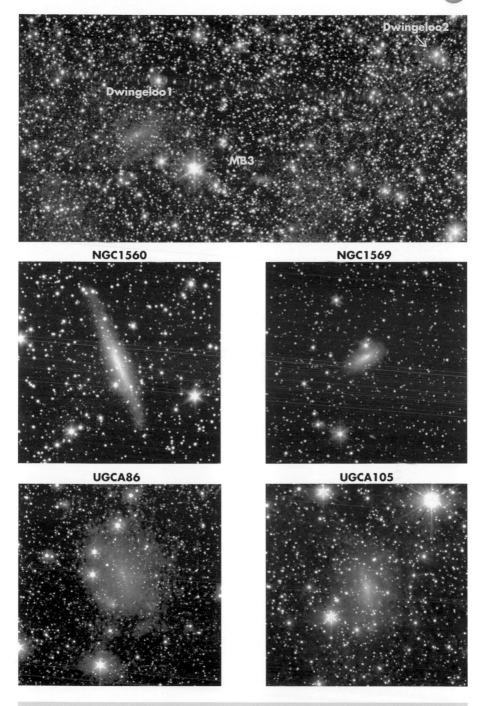

Figure 9.11. IC342/Maffei Galaxy Group montage (2). Red/infrared images with exposure totaling 45 minutes (90 × 30 seconds).

10 iterations – any more and noise tends to become objectionable. It then requires a "stretch" to make the faint detail visible. Logarithm is usually too strong and a gamma stretch with a value of 0.5 is generally about right. If I cannot decide which option is better (a frequent dilemma), then I sit on the fence and take an average of the DD and RL results!

Light pollution manifests itself by producing images that are grainy or speckled. This is the random element that cannot be removed. Now for that smarter processing – the magical noise-reduction filter. By *filter* I mean a digital process, not a special piece of glass. For this I have found IRIS's commands the best (Adaptive and Wavelet). These work well on diffuse nebula and elliptical galaxies where lack of smoothness in the image is all too obvious. They turn our grainy light-polluted images into those (almost) matching the smoothest ones from the darkest of sites (see Figure 9.7).

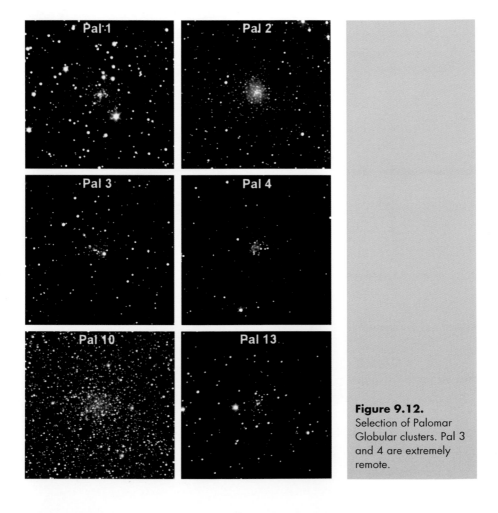

Figure 9.12.
Selection of Palomar Globular clusters. Pal 3 and 4 are extremely remote.

Color Imaging

Color imaging from a light-polluted environment presents many problems. We can no longer just image in the red as we need the full spectrum. My philosophy has always been to take the best mono image and then worry about the color later. As mentioned earlier, virtually all my images are shot through a red filter to reduce light pollution and increase contrast. This passes the red and infrared where light pollution is less prevalent. This has the added benefit, that when I want color, I already have a red frame. All that is needed is the blue one (with IR block). The green is the biggest problem, as this is where light pollution is worst. However, the green image can usually be synthesized by averaging the red and blue (see Figure 9.8). The blue image is inevitably light-polluted and much more noisy than the red. However, using the LRGB method, the red frame is used as both the R and L (luminance) frames, preserving the detail of the black-and-white image and only adding the color from the low-quality color image. MAXIM makes this fairly straightforward.

I have also used the color from my old photographic slides, having an extensive library of color images, which again can provide the low-resolution color information to supplement the high-resolution CCD image (see Figure 9.9). The final processing stage for color images is inevitably that noise-reduction filter again.

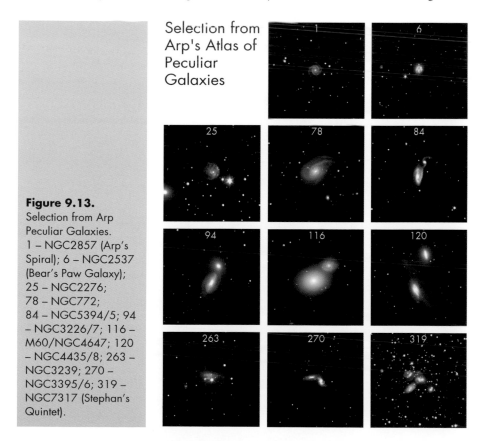

Selection from Arp's Atlas of Peculiar Galaxies

Figure 9.13.
Selection from Arp Peculiar Galaxies.
1 – NGC2857 (Arp's Spiral); 6 – NGC2537 (Bear's Paw Galaxy); 25 – NGC2276; 78 – NGC772; 84 – NGC5394/5; 94 – NGC3226/7; 116 – M60/NGC4647; 120 – NGC4435/8; 263 – NGC3239; 270 – NGC3395/6; 319 – NGC7317 (Stephan's Quintet).

Infrared Imaging

Because CCDs are sensitive to the (near) infrared (IR), imaging in this part of the spectrum is remarkably easy. I use a simple Hoya IR filter bought from a camera shop. This filter appears as black glass, as it transmits no visible light, making it the ultimate light-pollution filter! Amazingly it gives your telescope remarkable penetrating power. Galaxies hidden behind the Milky Way, such as the Maffei/IC342 Group and their faint companions, can be successfully recorded from light-polluted sites (Figures 9.10 and 9.11). Many of these are so faint they have only been recognized in the last 10 years – yet we can image them! The Quasar images (Figure 9.1) are another example – their high redshift carries their peak emission point (Lyman Alpha) into the near infrared – perfect for us! This is what I meant by being smarter in image choice.

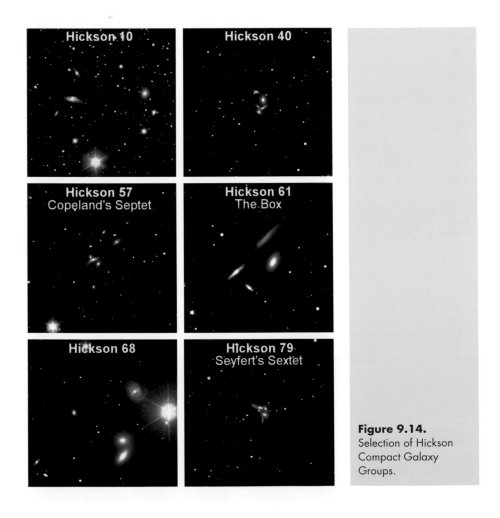

Figure 9.14.
Selection of Hickson Compact Galaxy Groups.

Figure 9.15. NGC2174 – mosaic of four sets of images and LRGB color. FLI Maxcam and 16-inch Newtonian.

Examples

The examples I have chosen reflect my interest in the obscure! Having lived through several major astronomical discoveries and controversies, I now find I can actually image many of the objects that featured in these debates. These include the faint Palomar Globulars (Figure 9.12), discovered by many famous astronomers using the Palomar Observatory Sky Survey (POSS), the Arp Peculiar galaxies (Figure 9.13) and the Hickson Compact Galaxy groups (Figure 9.14).

Conclusion

I hope that this has shown that there is plenty of scope for the suburban/urban amateur astronomer and that we can actually image many of the most interesting objects known. We just need to be a bit smarter. One final point: I would always recommend keeping records of what you did in taking the images and their processing. If something turns out great, you want to be able to repeat it!

Useful Web sites

Finger Lakes Instrumentation – CCD Camera Manufacturer
http://www.fli-cam.com/

Christian Buil's IRIS Software
http://www.astrosurf.com/buil/us/iris/iris.htm

Diffraction Limited's Maxim Software
http://www.cyanogen.com/products/maxim_main.htm

The IC 342/Maffei Group Revealed–Buta & McCall
http://www.journals.uchicago.edu/ApJ/journal/issues/ApJS/v124n1/39629/39629.text.html

Section 3

Advanced

CHAPTER TEN

The Hybrid Image: A New Astro-Imaging Philosophy

Robert Gendler

Introduction

This chapter is less about new methods or techniques in imaging and more about an entirely new philosophy for astro-imaging. Since my start in astrophotography, my goal has been to create high-resolution celestial portraits, with generous field size and maximal visual impact. I decided early on to explore the boundaries of amateur astro-imaging and that professional images would be my model for the quality I wanted to achieve. The difficult challenge would be to accomplish this with amateur equipment from a light-polluted location. This would require new and creative ways of putting images together.

Within the framework of this new astro-imaging philosophy, "image taking" grew into "imaging projects" and inevitably I found myself investing large amounts of time at the telescope gathering data and at the computer processing it.

The Hybrid Image: What Is It?

The philosophy of creating "hybrid images" evolved for me over the last several years after taking many images. According to Webster's dictionary, the generic definition of the term *hybrid* is "something of mixed origin or composition." A hybrid image is therefore defined as an image assembled from different components, in which each component of the image can have a different resolution or spectral properties. Hybrid images can range from the very simple to the very complex.

Hybrid images can be assembled using data from:

1. any combination of different focal lengths or f-ratios;
2. different telescopes, different detectors or different media and
3. conventional RGB filters, narrowband filters or both.

The essence of the hybrid image philosophy is that there is tremendous freedom and few rules. The only rule is that we acquire the data from the sky. We are then free to put the data together in any way we choose.

The doctrines of the hybrid image philosophy are the following:

1. to make amalgamations of different image data with the goal of producing a single unique image and
2. to allow the object and its environment to dictate the techniques and methods needed to record it in the most optimal way.

The Hybrid Image: Why Attempt It?

There are certain challenges unique to astrophotography. Astronomical objects have both large- and small-scale detail. They often have an extraordinary dynamic range with very dim and very bright areas in the same field. Many objects have emission properties not captured by conventional filters. Images have much greater impact if the object is viewed in its environment. Therefore a large field size is highly desirable. The last point is an important one as the most powerful astro-images evoke the vastness of our universe and possess a three-dimensional quality. This can only be achieved with highly resolved detail spread over a large field showing the object within its setting.

The technique of creating hybrid images answers these challenges in the following way. Small- and large-scale details are recorded using appropriate focal lengths – long focal length for small-scale detail, intermediate and short focal length for large-scale detail. We want to achieve large field size without sacrificing resolution, so mosaics play a key role. We capture both dim and bright areas with equal detail by using appropriate f-ratios and narrowband (hydrogen alpha) filters if needed. We then have the difficult task of composing all the above into a single unique image.

When I started astrophotography, one of the most difficult obstacles in creating these types of images was the inability to incorporate multiple frames, taken at different focal lengths, into the same image. In 2000 Auriga Imaging brought out the software program RegiStar. RegiStar essentially revolutionized amateur astro-imaging and provided the tools to go forward with creative and complex imaging projects. RegiStar allows the imager to register two or more images taken at different image scales so they can be used to create a single unique image (see Figure 10.1).

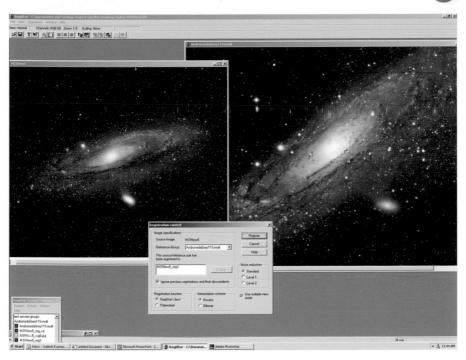

Figure 10.1. The working space of RegiStar allows for easy registration of frames taken at different focal lengths, for later compositing.

The Hybrid Image: How to Create It

Types of Hybrid Images

Mosaics. Why bother with mosaics if wide-field vistas can be imaged using short-focal-length instruments? The answer is that when using a shorter focal length to achieve greater field size there is an inevitable loss of resolution. Even if we use one of the newer-generation large-format CCD chips, there are few optical designs that have a sufficiently corrected field to take advantage of these large chips. Therefore there remains a very real role for the art of mosaics in producing large-format images while maintaining a high level of resolution.

Soon after I began astrophotography I became intrigued by the idea of splicing together multiple frames to achieve exciting large-format images. I started creating mosaics using a small refractor but, soon realized that, if I really wanted to explore the boundaries of amateur astro-imaging, I would need to use a longer-focal-length instrument. At this time I had purchased a 12.5-inch Ritchey-Chretien

Figure 10.2A. Grayscale mosaic.

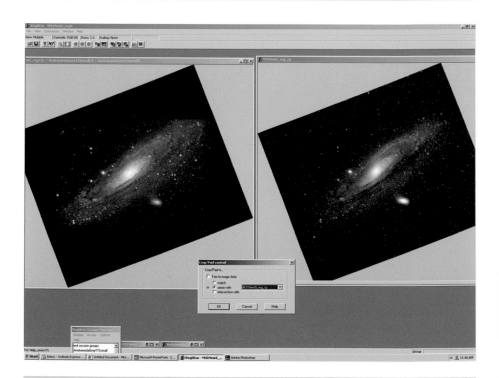

Figure 10.2B. Registering the color data with Registrar.

Figure 10.2C. This high-resolution mosaic of M31 required stitching together 40 separate frames over three months for a total cumulative exposure of 50 hours. To create a color version, color data were taken from an older, lower-resolution image of M31 and registered to the grayscale mosaic using RegiStar. An LRGB composite was then made in Photoshop using traditional methods. The grayscale mosaic was made using a 12.5-inch RC at F9. The color data were taken with an ST10E and 4-inch refractor at F5.

Cassegrain with a prime focus of 3000mm. Using this instrument I was able to produce mosaics of much higher resolution and much greater image scale (see Figure 10.2 a–c), but this also required much more work and time stitching the many frames together as the field of view was contracted. These are truly labor-intensive projects but the rewards are well worth the effort. The very large file sizes, generated by this process of combining multiple images, are sufficient for making very large-format prints. For me this is the most rewarding aspect of the process.

Probably the most critical element to success with large mosaics and other complex imaging projects is proper planning. I attribute any success I have had to proper planning. I often use programs like "The Sky" to help me plan the number and size of frames needed. I also make charts of the object using a template of the field of view of my imaging system. Preparation in this way is an essential exercise before going out with the telescope.

The art of creating mosaics can be found in detail on my Web site at http://www.robgendlerastropics.com/Article3.html and in *Sky Telescope*, June 1999, pp 138–141.

Composites. Composites are images made by layering separate frames taken of the same field using different techniques. Each frame is imaged using tech-

Figure 10.3A. High-resolution image of M81.

niques needed to optimize resolution and contrast for each part of the final image. The different frames are registered using RegiStar and then blended or layered together in Photoshop. The classic composite would be a short exposure of the trapezium layered over a deeper image of the Orion nebula. At about the same time I began creating mosaics, I also became interested in nontraditional composites. I asked myself why not substitute long-focal-length high-resolution images of selected areas, having abundant small-scale detail, onto a low-resolution background, which lacks small-scale detail and therefore does not require labor-intensive high-resolution imaging? In this way, the detail will be recorded where it is needed while other parts of the field that lacked small-scale detail would be spared the labor of high-resolution imaging. This makes for efficient but creative imaging (see Figure 10.3 a–c).

Hydrogen Alpha Color Composites.

A natural evolution of hydrogen alpha (H-alpha) imaging of nebulae is the creation of color images using narrowband H-alpha data. The primary obstacle to incorporating H-alpha data into a color image is the unnatural "fit" of the narrowband data into the luminance of a traditional LRGB composite. Initial attempts to use H-alpha data as luminance were disappointing and often resulted in muted colors, monochrome-appearing nebulosity and bizarre-looking stars.

Figure 10.3B. High-resolution image of M82.

Figure 10.3C. High-resolution images of M81 and M82 (taken at a focal length of 3 meters) were layered on a background taken at 1 meter FL to achieve the final composite. The high-resolution components (Figures 10.3A and B) were taken using a 12.5-inch RC at F9 and the background was taken with an STL11000 and AP155 at F7. (Note the small satellite galaxy near M81 is Holmberg IX.)

After quite a bit of experimenting I found that a natural-appearing image with aesthetic color balance could be obtained by incorporating the H-alpha data directly into the red channel of the RGB. I discovered that there were several ways to accomplish this successfully. One method was to make a 50/50 blend of the red channel and H-alpha data in Photoshop using the normal blending mode. The resulting image had much of the desirable contrast and signal of the H-alpha data along with the traditional-appearing stars of the red channel. I later modified this

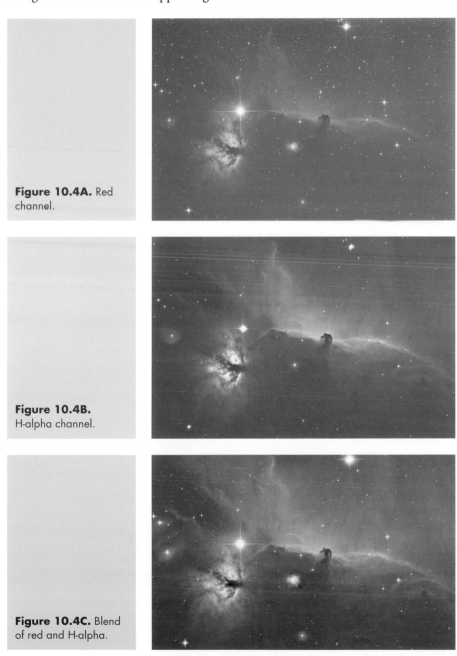

Figure 10.4A. Red channel.

Figure 10.4B. H-alpha channel.

Figure 10.4C. Blend of red and H-alpha.

Figure 10.4D. To make an H-alpha color composite of the Horsehead region I first created a blend of the red-filtered exposures and hydrogen alpha–filtered data using normal mode in Photoshop. After flattening this blend, I had a new red channel with which I made a traditional RGB. Cumulative exposure time was 12 hours. The background was taken using an AP155 refractor at F7. The high-resolution components were taken with a 12.5-inch RC at F9.

Figure 10.5A.
Grayscale mosaic.

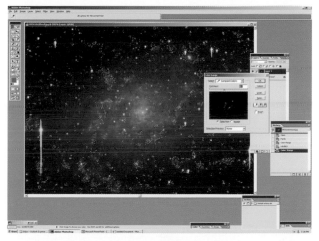

Figure 10.5B.
Low-resolution RGB
image.

Figure 10.5C.
H-alpha image.

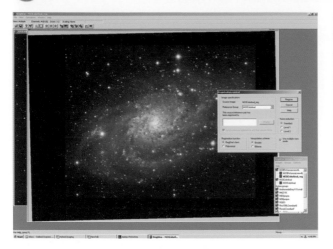

Figure 10.5D.
Registration of
component images
using Registrar.

technique so that the H-alpha data can be used at 100% opacity in the blend. Using the red channel as the top layer one can then substitute the stars of the red channel into the H-alpha layer using the lighten mode at 100% opacity. Alternatively, if the previous method doesn't work well, the stars in the red channel can be selected using the color range (choose *Highlights*, then expand 1 to 3 pixels), inverse select and delete everything but the stars. Once the image is flattened it will have all the signal of the H-alpha component and the stars of the red channel. It may also help to first register the red channel with the H-alpha image using RegiStar since the data is often recorded on different nights and may need aligning. Once the new red channel is prepared it can be combined with the green and blue channels in a traditional way. Because of the intensity of the H-alpha/red channel the final RGB may need to have the "red" adjusted down. If needed, the H-alpha/red channel can be used as luminance in a traditional LRGB if this further enhances the image.

The final H-alpha color composite will have the remarkable depth of the hydrogen alpha image plus the beauty of a well-balanced color image (see Figure 10.4a–d).

Complex Hybrid Images

The natural evolution of these techniques is the construction of very complex images making use of all the methods described. Images of this nature are complex amalgamations of different resolution and color components and are often mosaics.

These types of imaging projects often require many hours of data collection and can take weeks or months to complete. The time and energy spent are almost always well worth it as the finished product is often a very unique and striking image and can be printed at tremendous sizes. (See Figure 10.5a–e and Figure 10.6.)

Figure 10.5E. For M33 a 15-frame grayscale mosaic was created first. This was done at long focal length (3000mm) using a 12.5-inch RC and ST10XME to produce a high-resolution image with FWHM of 2 arcseconds or less. A low resolution RGB image taken with a 4-inch refractor was used for the color component. A separate color image was made using three hours of hydrogen alpha–filtered data to highlight the numerous HII regions present in M33. A blend was then made of the conventional RGB and H-alpha color composite using the "Color range" tool in Photoshop. The HII regions were selected from the H-alpha color composite and layered into the conventional RGB. Next the color image was registered to the high-resolution grayscale mosaic in RegiStar. The final image is a high-resolution color mosaic highlighting the numerous HII regions of M33. Cumulative exposure time was 21 hours.

Figure 10.6. The Orion deep field is a mosaic of four separate frames. Each frame is a hydrogen alpha color composite with layered high-resolution components. The background image data were taken with a 4-inch refractor at F5 and STL11000 camera. The higher-resolution components were taken with an ST10XME, AP155 and 12.5-inch RC. Cumulative exposure time was 20 hours.

Figure 10.7. Robert Gendler and his "observatory." Actually the telescope and mount are attached to a set of "wheelybars" (JMI), which is all wheeled out of the garage onto the driveway. The driveway is where all the imaging is carried out. This Is a typical night with the southern sky filled with the "skyglow" of adjacent towns and neighborhood lights. The limiting visual magnitude on an excellent night is about 4.5 magnitude.

Conclusion

Creating hybrid images is for those who enjoy a challenge and want to explore the limits of what can be accomplished using amateur equipment (see Figure 10.7). Though these types of images are often very labor-intensive and require substantial time and planning, the completed images can make spectacular prints and narrow the gap between professional and amateur image quality.

Useful Web site

RegiStar Software:
http://www.aurigaimaging.com/

CHAPTER ELEVEN

Amateur Spectroscopy in the 21st Century

Dale E. Mais

Introduction

Spectroscopy had its beginnings in the second half of the 19th century, when it was primarily the domain of the amateur working in his own private observatory. As we entered the 20th century with a greater emphasis on astrophysics, this area of research shifted toward the professional astronomer working in world-class observatories. This shift was primarily driven by the increasing costs and skills required to do state-of-the-art spectroscopy and the requirement for large telescopes due to film-based detection of a spectrum. Once again, we are experiencing a shift where the amateur can make contributions to the area of spectroscopy. This is due to both the use of more sensitive CCD detectors and the recent availability of powerful and versatile spectrometers aimed at the amateur community. I will focus on the instrument produced by Santa Barbara Instrument Group (SBIG), the self-guided spectrometer (SGS), and will provide the reader with an overview of the kinds of information an amateur can now obtain from his own backyard.

Equipment and Calibration

My primary instrument for spectroscopy and the evaluation of the SGS is a Celestron 14. The spectrometer is linked to the telescope with a focal reducer, giving a final f6 ratio. The CCD camera attached to the spectrometer is the SBIG ST-8E with 9-μm pixel size. The SGS instrument appeared on the market during the second half of 1999 and was aimed at a subgroup of science-oriented

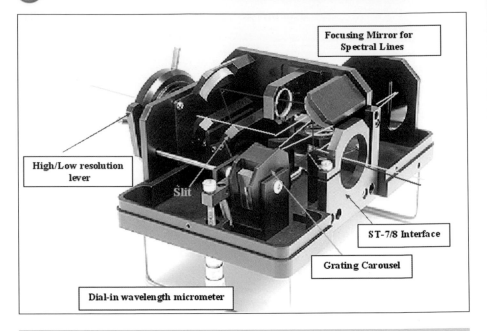

Figure 11.1. The Santa Barbara Instrument Group self-guiding spectrometer showing various features of the instrument and the optical path. The yellow (lighter) path shows the route followed by light passing through the slit to the grating and ultimately to the imaging chip of the ST-7/8 camera. The red (heavier line) path is the route followed by the light not passing through the slit and ending at the guiding chip of the same camera. This allows for extended guiding of a star. Either a different star not passing through the slit or residual light of a star on the slit can be used for guiding purposes. The micrometer screw controls the spectrum region to be observed and has been modified by attaching a JMI digital focusing motor. This permits the spectral region being examined to be controlled by software on a computer.

amateurs with special interest in the field of spectroscopy [1]. The instrument is shown in Figure 11.1 with a number of features pointed out and the path of light indicated. The instrument features several novel features. In conjunction with SBIG CCD camera's, the SGS is self-guiding in that it keeps the image of an object locked onto the entrance slit, which allows for long exposures to be taken. The light from the telescope reaches the entrance slit, which can be 18- or 72 μm wide. The light passes through the slit and reaches the grating and ultimately the CCD camera's imaging chip. The remaining field of view is observed on the guiding CCD chip of the camera and allows the viewer to select a field star to guide once the object of interest is centered on the slit. In this chapter, only results obtained using the 18 μm slit will be presented. The wider slit option allows the spectra of fainter objects to be obtained at the expense of resolution. This would be particularly useful to those interested in measuring the redshifts of more distant, and thus fainter objects, since the wider slit permits an additional 2 magnitudes of penetration. This is, however, at the expense of resolution.

The SGS features a dual grating carousel, which, with the flip of a lever, allows dispersions in both the low-resolution mode (~4 Angstroms/pixel, ~400 Angstroms/mm, 150 line grating) and higher resolution mode (~1 Angstrom/pixel, ~100 Angstroms/mm, 600 line grating). In the low-resolution mode, about 3000 Angstrom coverage is obtained, whereas in the high-resolution mode, about 750 Angstroms for the ST-7 camera and twice this for the larger ST-8. More recently, grating carousels with even higher dispersions have become available (0.5 and 0.3 Angstrom/pixel, 1200- and 1800-line gratings, respectively). The particular region of the spectrum is determined by a micrometer dial and is set by the user. The overall wavelength range of the unit is from approximately 10,000 to 3500 Angstroms. Spectra are obtained using any of the usual camera control software packages such as CCDSOFT, MAXIM or CCDOPS and analyzed using the software package SPECTRA (from SBIG) or VSpec, a free spectroscopy package [2]. Wavelength calibration was carried out using emission lines from a thorium/argon lamp. This type of emission lamp is widely used among professionals because it produces many lines across the visible spectrum. Figure 11.2 shows the author's spectroscopy setup with appropriate features labeled.

Figure 11.2. The author's spectroscopy setup showing the spectrometer-camera attached to the back of a C14. The thorium/argon lamp is used for wavelength calibration of the spectra obtained. The light from the lamp is introduced into a window on the bottom of the spectrometer. Thorium/argon lamps produce many lines throughout the visible and near infrared, allowing for easy selection of lines of known wavelength for the calibration steps.

Calibration of spectra initially follows the usual reductions done for images. However, once an image has been dark-subtracted, spectra require their own set of calibrations to be conducted. These include wavelength and flux calibration. The flux calibration step can be looked upon as being similar to flat-fielding where you manipulate your spectrum with a spectrum you obtained of a standard star (usually Vega or another B- or A type star). This is all accomplished with Vspec software relatively easily or IRAF, the professional standard that is much more difficult to use. A calibrated text file can be saved and imported into any of a variety of graphic spreadsheets, including Excel. Once the spectrum is expanded into a line profile, various routines can be applied that enhance the lines and make them easier to identify. Absorption and emission line identifications were carried out using tables from the *Handbook of Chemistry and Physics* [3] or, again using Vspec, which has an extensive line database.

Results and Discussion

The low-resolution mode is useful for stellar classification and obtaining spectra of planetary nebula. In the high-resolution mode, many absorption lines are visible of atoms, ions and simple molecules. Figure 11.3 shows the higher-resolu-

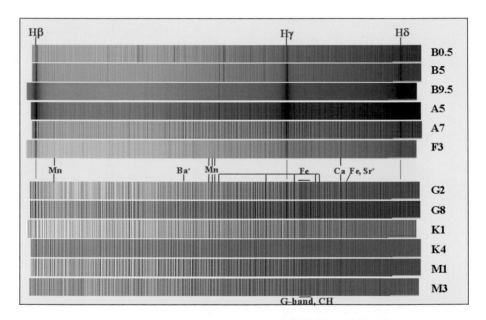

Figure 11.3. Classification of stars based on their spectra. Spectra were obtained in the high-resolution mode from Hβ to Hδ for stars from early B to M class. Note the increase followed by decrease in the hydrogen absorption lines as one proceeds from B- to M-type stars along with the general increase in the number of metal lines as the temperature decreases toward M-class stars. Several different metal lines are identified, along with the G band representing the diatomic molecule CH.

tion spectra of stars from class B to M and luminosity class III and spans the region from Hβ to about 4100 A. Several of the more prominent lines are labeled. As one can see, many lines are present, especially as one proceeds to cooler stars. Graphic display of these spectra, for example, using Excel, results in line profiles, especially for the more intense lines such as those for hydrogen. These profiles contain information regarding the physics of the stellar atmosphere. For example, buried in the line profile and intensity are such information as pressure, density, abundances and temperature, some of which, with the appropriate mathematics or software tools, can be extracted. In addition, rotation of the star is also contained within the profile, which also can be extracted [4].

Figure 11.4 shows the emission line spectrum of NGC7009, the Saturn Nebula, in the lower-resolution mode. The upper-left corner shows an image of the nebula with the slit superimposed over the nebula. The main body of the line spectrum was obtained in the low-resolution mode so that a majority of the emission features could be observed in a single spectrum. A higher-resolution spectrum is shown in the upper right, centered at the Hα line. In high-resolution mode, three components are seen to make up this strong emission, Hα and two lines of N^+, which flank Hα. A graphical presentation of the line spectra is also shown. The usual types of species are apparent in the spectrum of a planetary nebula. The hydrogen and helium lines, along with the prominent forbidden O^{+2} lines, Ar^{+2}, Ar^{+3}, N^{+2} and S^+, round out the variety of ionic species present. For extended objects, such as nebulae, one obtains a spectrum across the entire length of the slit. During the analysis of the results, the software allows you to select small subregions of the spectrum for analysis. As a result, one can profile the composition across the entire slit, effectively obtaining many spectra of the object at a single time, which is analyzed within the software. As discussed later, this would also allow for profiling temperature and density parameters of the nebula across the slit.

In addition to identifying emission features, other physical features of a nebula can be determined, such as the temperature (T) and electron density (N_e). Theoretically derived equations [5] have been obtained that relate electron density and temperature to line ratio intensities, as seen in Equations (1) and (2).

$$(I_{4959} + I_{5007})/I_{4363} = [7.15/(1 + 0.00028N_e/T^{1/2})]10^{14300/T} \tag{1}$$
$$(I_{6548} + I_{6584})/I_{5755} = [8.50/(1 + 0.00290N_e/T^{1/2})]10^{10800/T} \tag{2}$$

Equation (1) relates the line-intensity ratios for transition occurring at 4959, 5007 and 4363 Angstroms for O^{+2}, while Equation (2) relates the lines at 6548, 6584 and 5755 Angstroms for N^+. The bottom panel of Figure 11.4 shows the graphic solution of these equations, yielding a temperature and electron density. This example portrays the gold mine of information contained within a spectrum. As was mentioned earlier, for an extended object the spectrum across the entire slit is obtained and line ratios as a function of distance from the central star, for example, can be obtained in a single spectrum. The potential is present to create two-dimensional profiles of temperature, density and composition for nebular-type objects.

When the spectrometer is used in high-resolution mode, many absorption features can be observed in the spectra, particularly in cooler stars. Simple image-processing techniques enhance these features making their identification much easier. Figure 11.5 shows the high-resolution spectrum of 13-theta Cygnus, an

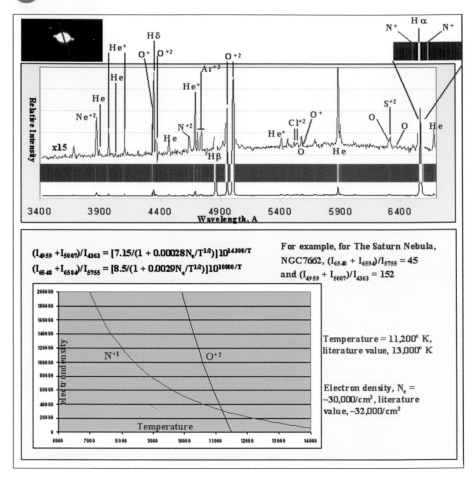

Figure 11.4. Spectrum of planetary nebula NGC 7009 (Saturn Nebula). In the upper-left corner of the top panel, the positioning of the slit is indicated. The low-resolution spectrum is shown as both a graph and an emission line profile. The high-resolution spectrum is shown in the upper right, centered at the Hα line showing the presence of N+ lines flanking the Hα line. Various other ionic and atomic species are identified. Many of these lines are forbidden, such as the intense 5007 Angstrom line of O^{+2}. The bottom panel shows the kind of physics that can be derived from such a spectrum. Theoretical equations describe the relationships between electron density and temperature to line intensity ratios. Two such equations are shown (there are many). The ratios are determined from the calibrated spectra and the two unknowns solved to yield electron density and temperature.

F4-type star, demonstrating the variety of elements, which can easily be identified. A few of the many lines have been identified and are labeled in the spectrum. The upper spectrum of the top panel identifies unionized elements in the spectrum, which spans the blue part of the visible spectrum from Hβ to Hδ, The lower spectrum of the top panel identifies ionized elements and covers the

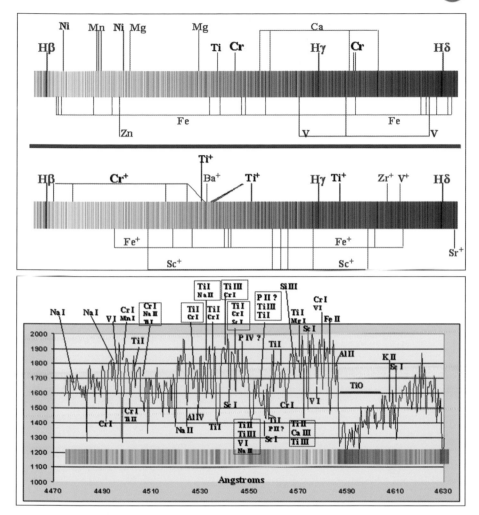

Figure 11.5. The upper panel shows the spectrum of 13-theta Cygnus and an F-type star in the blue region. The upper spectrum of that panel identifies a variety of atomic species, while the lower panel shows the same spectrum but identifies ionic species. The presence of absorption lines, mostly of metals, is even more pronounced in cooler M-type stars as shown in the lower panel. This is a spectrum of omicron Ceti (Mira) near maximum light and shows the great number of lines, many of which are blends of several different lines. This spectrum spans only 160 Angstroms and is the result of using a higher-resolution grating than the one used in the upper panel.

same wavelength region. Even more dramatic are the lines present in cooler stars such as are seen in the lower panel of Figure 11.5, which is a spectrum in higher resolution (0.5 Angstrom/pixel) of omicron Ceti (Mira), an M-type star. At this point let us diverge a bit and talk about lines.

The identification of species responsible for the observed absorption lines remains somewhat of an art, since no software packages are available that will do this. The *Chemistry and Physics Handbook* lists more than 21,000 lines between wavelengths of 3600 and 10,000 Angstroms for all the elements. This represents, on average, about three lines per Angstrom wavelength interval. Since the instrument is only at best able to resolve 0.3–0.5 Angstrom in the highest resolution mode, some criteria must be used to eliminate many of the potential lines observed or, as is often the case, a line may represent a blend of two or more lines. The presence and intensity of a feature, due to an atom or ionic species, is the result of many parameters such as natural abundance, probability of the absorption, temperature, density and pressure. For our purposes, the first three are the most important. Many potential features can be eliminated at the start simply because the natural abundance of an element is so low. For example, at a given temperature the $H\beta$ absorption line at 4861.33 Angstroms has a relative intensity of 80 compared to 3000 for protactinium at 4861.49 Angstroms. Yet one would not typically assign this absorption line in a stellar spectrum to protactinium, simply because the natural abundance of this element is 9 orders of magnitude lower than that of hydrogen. In addition to this abundance factor, other absorption lines for hydrogen are present where they should be, whereas protactinium lines are not. To be certain of assignment of a line, it is important to find several lines for a particular species. In addition, for most of the spectra shown in this chapter, the spectra obtained were compared directly with those obtained in the astronomical literature to be certain of the assignments. My initial goals in using this instrument were to establish how far the instrument could be pushed in giving spectra and in the resulting identification of features.

Figure 11.6 illustrates just how far this instrument is capable of being pushed to give useful data. 19 Piscium is a carbon-type Mira variable star. These type of stars are my particular interest and represent old, evolving stars nearing their deaths. Stars of this spectral class often exhibit absorption lines owing to the unstable element technetium. This element is not found naturally in the solar system because its longest lived isotope ^{97}Te, has a half-life of only 2.6×10^6 years and, as a result, all technetium endogenous to the formation of the solar system has long since decayed. Yet multiple lines of technetium can be detected in the atmosphere of this star and others of the S and C types [6]. It is known that at certain stages in the evolution of these stars neutron capture reactions, on seed iron-nickel atoms, are proceeding deep within stars of this type. At other stages in the stars' evolution, dredging mechanisms bring to the surface these nuclear-processed materials, now well laced with heavy elements including technetium. These heavy elements such as zirconium and zirconium oxide molecules, now often appear in the stars' atmospheres, which give rise to extensive banded structure in the spectrum in the red region.

The other very interesting aspect of this type of star is the fact that they often contain abnormal amounts of ^{13}C compared to ^{12}C. The normal solar system ratio of ^{12}C to ^{13}C is 80, but in many of these type of stars this ratio can approach 4. This follows from theoretical work, which suggests that the neutron source in these stars, which gives rise to neutron capture heavy elements, is due to the burning of pockets of ^{13}C deep down in the star. The dredge-up results in the presence of these heavy elements in the stellar atmosphere and can result in

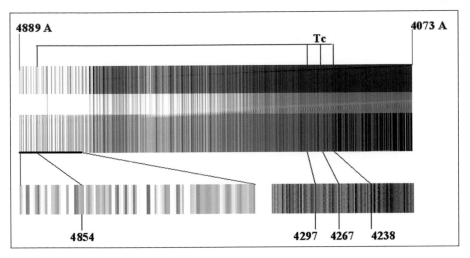

Figure 11.6. The presence of the unstable element technetium is shown in the carbon star 19 Piscium. Many stars of the S and C types exhibit these lines and indicate that heavy element synthesis, via slow neutron capture processes, are occurring in these evolved stars and being dredged up to appear in the surface layers. Since technetium is an unstable element with a half-life of only a few million years, this indicates that heavy element synthesis is an ongoing process, not limited to supernova explosions. Many other heavy elements are often also visible such as strontium and zirconium.

enhanced quantities of ^{13}C. Normally, the detection of isotopes of atoms or ions cannot be done easily because the lines are extremely close together and normal line-broadening effects cause them to overlap. However, this is not the case with molecules where rotation and vibration transitions are much more sensitive to the isotope composition. These types of transitions are normally outside the optical range, but electronic transitions can couple with vibration transitions giving rise to a blanketing effect of the multitude of lines that result, and these lines are often observable in the optical region. In cooler S-and C-type stars, diatomic carbon forms and a clear separation of absorption lines occurs between diatomic ^{12}C-^{12}C, ^{12}C-^{13}C and ^{13}C-^{13}C as indicated on the spectrum in Figure 11.7.

Figure 11.7 shows the graphical spectra of three different stars containing varying ratios of ^{12}C to ^{13}C. Some of these stars are noted for their large quantities of carbon as observed with diatomic carbon. All three possible combinations of diatomic C-C (soot) are observed, ^{12}C-^{12}C, ^{12}C-^{13}C and ^{13}C-^{13}C. Spectrum **A** represents a carbon star with a solar system–like ratio of the two isotopes (^{12}C/^{13}C of around 80). In spectrum **B** this ratio has decreased to 20 and results in the appearance of ^{12}C-^{13}C in the spectrum. Finally, in spectrum **C** the ratio is around 4, and ^{13}C-^{13}C makes its presence known in the spectrum. The blanketing effect of diatomic carbon is shown along with the identification of several metal lines.

In addition, with careful wavelength calibration, one can measure the Doppler shift of absorption and emission lines to determine velocities of approach or recession of objects along with rotation velocities of stars and planets. I have been

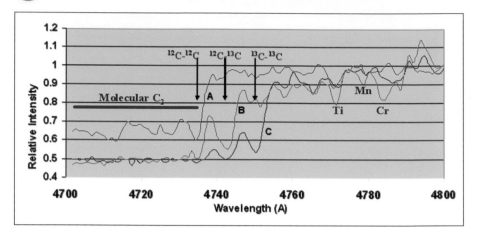

Figure 11.7. The spectra of three different carbon stars (C stars) are shown in this graph in the 4700–4800 Angstrom region of the spectrum. Proceeding from **A** to **B** to **C**, the $^{12}C/^{13}C$ ratio decreases from 80 to 20 to 4. This results in the gradual appearance of diatomic carbon (C_2) containing various combinations of carbon isotopes. Note the large blanketing effect C_2 has on the spectrum blueward of 4740 Angstroms. This can be so dramatic in these type of stars that the blue part of the spectrum can essentially disappear.

able to successfully observe and measure the rotation of Saturn by aligning the slit along the rings of the planet. In only a few-second-long exposure one is able to obtain a spectrum of the entire disk and ring system of Saturn. The eastern and western region of the image can be isolated using SPECTRA software, and each of the spectra calibrated against the $H\alpha$ line. When the final two images are aligned, they show the clear shift in the lines, which occurs due to Saturn's rotation. From this the rotation velocity can be calculated. In a similar manner, the velocity of approach of M31 has been determined by acquiring the spectrum of the nucleus of the galaxy. In this case, only the narrow 18-μm slit was used, and as a result a relatively long 60-minute exposure was necessary. I have not tried using the wider slit, but according to the instrument specifications, a gain of 2 magnitudes is possible, with a loss of resolution, which would occur with the wider slit, but one should be able to determine the center of lines with the software provided.

Future Directions

The future is certainly bright for amateur spectroscopy. As far as the SGS is concerned, grating carousels are now available that contain a 1200/1800 line grating combination, providing greater dispersion of the spectrum. The unit comes standard with a 150/600 line grating combination. This definitely improves the identification of lines by spreading the spectrum out by a factor of two to three compared to the 600-line grating. Of course, this comes at the expense of sensitivity

Figure 11.8. Dale Mais in his observatory, only 12 miles from Mount Palomar.

since you are spreading the light out more. In addition, user-friendly and versatile software for the amateur spectroscopist would be most welcome. While IRAF is the current standard for this type of work among professionals, its requirement of a UNIX or Linux operating system, along with what I understand to be a difficult package to learn at best, will make this software little used by the budding spectroscopist. The package VSpec definitely fills a void in this regard. It is Windows-based and very powerful, and nearly all amateurs involved in spectroscopy use it.

Spectroscopy is much more involved than other areas, such as photometry, where amateurs are making scientific contributions. It is relatively early in the amateur spectroscopy arena and how this field will develop remains to be seen. One thing is for certain: amateurs will not make contributions in this area unless they do so in collaboration with a professional. To this end, for those interested, I suggest two different groups where you can learn more:

1. The Society for Astronomical Sciences, http://socastrosci.org/Default.htm
2. The Working Group for Professional-Amateur Collaborations, a committee of the American Astronomical Society, http://www.aas.org/wgpac/.

Conclusions

The Santa Barbara Instrument Group Spectrometer represents a quantum leap forward for the amateur interested in the fertile field of spectroscopy. Even with a relatively small telescope, this instrument coupled to sensitive CCD cameras and utilizing the self-guiding feature of the ST-7/8 camera allows one to reach unprecedented magnitudes. Spectral analyses only dreamed of by the amateur a few years ago can now be carried out. Even after using the instrument for several years, I remain astounded by the fact that an amateur, with only relatively modest equipment from his own backyard, can detect technetium, many dozens of other elements, simple molecules and carbon isotopes in stars or nebulae hundreds of light years away.

References

1. SBIG Web page, www.sbig.com.
2. Valerie Desnoux, http://astrosurf.com/vdesnoux/.
3. *Handbook of Chemistry and Physics*, 79th edition, 1998–1999, section 10-1 to 10-88.
4. *Optical Astronomical Spectroscopy*, C.R. Kitchin, Institute of Physics Publishing Ltd, 1995.
5. *Astrophysical Formula*, K.R. Lang, Springer-Verlag, 2nd edition, 1980.
6. *The Classification of Stars*, C. Jaschek and M. Jaschek, Cambridge University Press, 1987.

CHAPTER TWELVE

Successful Patrolling for Supernovae

Tom Boles

Introduction

Patrolling for supernovae is not difficult. All that is required is the ability to take quality images of galaxies very quickly and compare them with a library of comparison images. This simple statement will dictate how the process must be executed. The key requirement here is speed, the ability to record hundreds of images per clear night. To do this successfully requires not only suitable software but also a mount, which is able to point consistently at selected galaxies time after time and, at least, position them somewhere within the area of the chip. This dictates the pointing quality of the mount and the field size of the camera.

The Hardware

I have chosen a combination of mount and software to do just that. Both of these come from Software Bisque. This ensures ongoing compatibility between the mount and its control program. The mount is the Paramount. This is widely recognized as the most accurate commercial mount available. I own three of these, two model 1100s and one 1100S. These are early versions of the Paramount ME, that, with its enhancements, is even more accurate. My mounts and control software are capable of pointing with an accuracy of less than one arc-minute across the entire sky. The focal length of each of my three C14 f/11 SCTs at 154 inches gives me a field of just under 11×11 arcminutes. This is big enough to capture all but the largest galaxies and makes it easier to ensure the galaxy is found. It allows for catalog errors that can sometimes cause incorrect locating of galaxies.

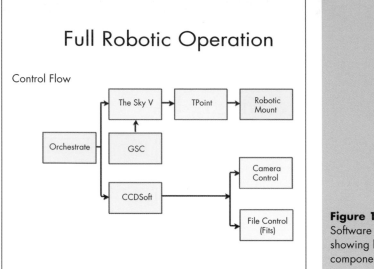

Figure 12.1.
Software control flow, showing how all the components fit together.

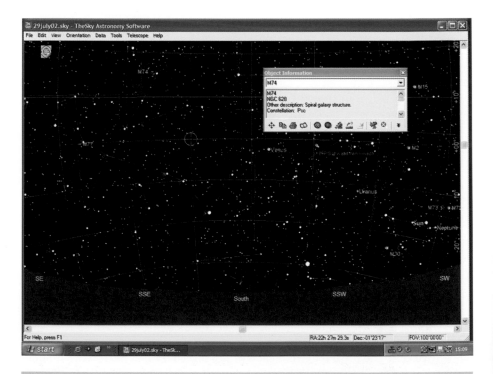

Figure 12.2. Screen shot of TheSky, the main control software for the mount.

The Software

The software falls into three categories: the camera control, the mount control and the scripting programs. In addition to the control software is a program that enhances the pointing accuracy of the mount even further. See Figure 12.1 for a block diagram of how the software programs work together.

The main control program for the mount is TheSky level 5. This uses the GSC catalog and can send coordinates to the mount for slewing to any chosen galaxy in the catalog. The instruction can be given manually by clicking on a galaxy shown on the screen, by inputting one of its names and clicking on the slew button or by entering its coordinates directly (see Fig. 12.2).

This would be a rather slow process if a large number of galaxies were to be patrolled and all but impossible if more than one telescope is to be controlled. For this reason, a scripting program is available that will input the galaxy names or coordinates. Called Orchestrate, this is also part of the Software Bisque suite. Orchestrate is capable of reading a galaxy's name or coordinates from a script and sending the slew instruction to the telescope. After allowing time for it to settle (which is also programmable), it can control the camera's exposure via yet another program in the suite called CCDSoft. See Figure 12.3, which shows the Orchestrate program screen opened with a sample script.

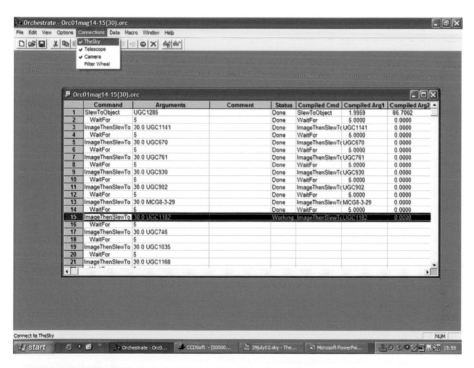

Figure 12.3. Screen shot of Orchestrate, a scripting program for controlling the telescope and associated software.

Figure 12.4. Screen shot of CCDSoft, which is used for controlling the camera.

CCDSoft automatically triggers the camera shutter, takes a dark frame if so instructed and stores the image taken with a title in a directory whose name is preselected by the user. For the purposes of supernova patrolling, a dark frame is necessary to eliminate hot pixels, which can obscure candidates. I do not bother with flat fields unless accurate photometry is required. This saves time and speeds the process. Figure 12.4 shows the screen of CCDSoft and the progress bar showing the time into the current integration.

The final but still important program is TPoint. This is a clever program that improves the pointing accuracy of the telescope significantly. The user has to train TPoint in advance by slewing the telescope to known locations, usually stars. The star is then centered on the screen. The difference between where the telescope was pointing and where it should have pointed is a measured error. The program remembers this for locations covering the sky and reproduces a correction whenever the telescope slews in future. The software uses correction terms that interpolate the errors allowing any location in the sky to be corrected. The more points the user trains, the more accurate the slewing becomes, up to a point of minimal returns. This is usually around 100 points. More can be used for training if long-exposure, unguided photography is required. The Paramount ME uses a feature called Protrack, which continually corrects guiding using the TPoint corrections during long exposures. Effects due to mount flexure and atmospheric refraction are therefore reduced. Figure 12.5 shows TPoint opened and a scatter diagram, which displays the corrections that it has calculated for different areas

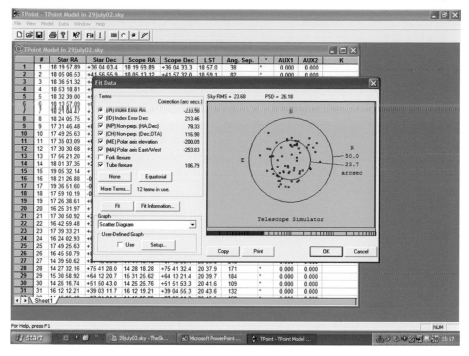

Figure 12.5. Screen shot of TPoint. Used to improve the pointing accuracy of the telescope.

of the sky. This can be used to tune the operation and balance the mount. It is also a very good tool for accurate polar alignment (see Fig. 12.6). All the data shown here are simulated.

This might sound a little complicated. I can assure you that it isn't. Once the software is loaded, much of its use is intuitive and familiarity will come very quickly. The result will be a system capable of slewing and pointing with an accuracy of better than one arc-minute. It will do this completely unsupervised if you are brave enough to leave it. I prefer to keep sight of what each of the systems is doing using a local area network and a program like pcAnywhere from Symantec. Windows XP Professional has a similar feature included as part of the operating system. I must admit I have left it for up to three hours, completely unsupervised, while visiting my local pub. It is only seven minutes walking distance from the observatory so it didn't require too much courage, and I could easily sprint back quickly if the weather changed.

Now the Hard Work Begins

So far I have described the programming and acquisition of images. Up to now, other than creating the scripts for Orchestrate, which can be saved and reused,

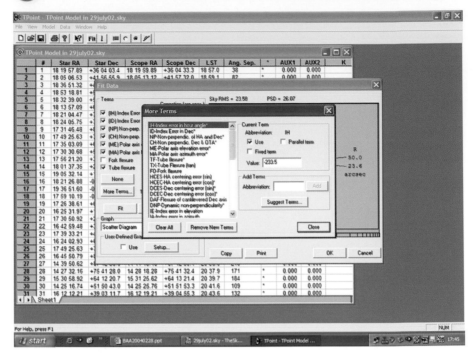

Figure 12.6. TPoint showing more control terms.

there is not a very large workload for the observer. I switch on at the beginning of the evening, take the cameras down to their operating temperature, focus, load the first script and off I go. The systems do most of the work. With this setup I can use either 30- or 60-second integration times. I usually opt for 60 seconds, but on a particularly clear night with good seeing I might go for 30 seconds and still reach a little short of 19th magnitude. At 30 seconds integration and allowing for settling time, I can acquire 210 images an hour using the three telescopes. In the UK in November, darkness can last for 14 hours or more. At 210 images an hour that is just under 3000 images in an evening. In practice, this is never achieved because a previous night's suspects have to be rechecked and there are pauses to change scripts or refocus. It is possible to regularly get more than 2400 images at the right time of year. This is where the real hard work begins. There is no available software that allows amateurs to automatically check images, that is both quick enough and allows the images to be checked easily at all the necessary histogram settings. A full range of stretch settings is necessary to pick up the fainter supernovae in the outer arms and in the bright central portions close to the galaxy's bulge. The only way that I have found to be reliable and quick enough is to manually compare a pair of images in CCDSoft (see Fig. 12.7).

This shows two images opened simultaneously, in the same window, ready for comparison. One is the master, the other a recently acquired patrol image. Thankfully, Windows allows multiple images to be opened simultaneously from

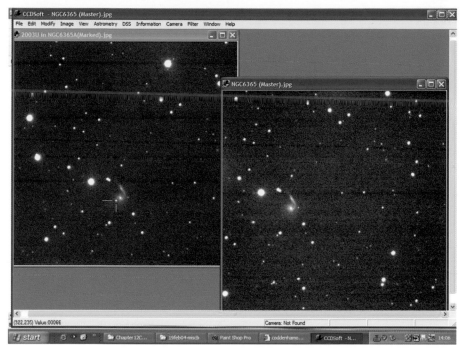

Figure 12.7. CCDSoft with supernova SN2003u (marked) and comparison master image.

within Windows Explorer. I usually open the patrol images, up to 50 at a time, and then selectively open masters one by one. This allows the image settings to be changed to display the inner and outer parts of the galaxy image alternately. As well as looking for possible supernovae, I compare the new image just taken with the older master image. If it is of higher quality, I replace it. The master image stored on the hard drive is therefore the best image that I have ever taken of that particular galaxy. It is essential to use your own master images. They match your telescope setup and the color response of your camera. This saves a lot of time and helps you to dismiss unlikely candidates quicker and recognize potential candidates more easily. By all means, check your images against the Palomar Sky Survey, particularly the second-generation plates. Indeed, the Central Bureau (CBAT) will require you to have done this before submitting your report on a suspect. This is to support your claim by demonstrating that the star is not visible on the older images, but it is no substitute for good-quality masters taken on your own system. The process of image checking can take up to 40 seconds per image. That means that I can just about keep up with one telescope. If there is a string of clear nights, which is usual, I can have thousands of images in the backlog waiting to be checked. This can take up to a week to clear. I rely on the predictable British weather to cloud over for days to let me recover: it rarely lets me down.

Some Tips and Some Traps

From this point on, past experience of using CCDs helps enormously. After you have studied several hundred images, it is possible to get to know your camera pretty intimately. You recognize where on the chip there are likely to be hot pixels and blemishes and how it will perform in various sky conditions. Variations in transparency can be the greatest. Surprisingly, and up to a point, small amounts of mist can assist in achieving fainter magnitudes. The increased stability that it causes results in pinpoint stars, concentrating the starlight on a smaller area of the chip. Hot pixels regularly appear. Knowledge of the chip temperature and a pixel's location can often allow you to eliminate it as a supernova candidate. With three cameras this becomes increasingly difficult. The usual clue to the existence of a hot pixel is the lack of a point-spread function (PSF). Moving the histogram settings should cause a true PSF to increase and decrease in diameter. You should nevertheless be aware that true stars can look very pixular on those very good nights. Stars falling on gaps between pixels can also have odd appearances. Elimination as early as possible in the process is the key to success. This is because even after a candidate is suspected, a lot of checking is required. Your most valuable asset is telescope time under suitable skies. I do all other checks before I consider stopping a running script to test for a suspect. An excep-

Figure 12.8. Tom Boles with two Paramounts in the roll-off roof observatory.

tion to this might be if it were brighter than magnitude 15 and very far from the ecliptic. This leads nicely into a list of checks and the process that I routinely make.

The Process and Precautions

The number of supernovae discovered in recent years has increased dramatically due to the availability of better hardware and applications. For this reason I have recently moved the checking of the Central Bureau's (CBAT's) "Recent Supernova" Web page to the top of the list. I check this, even though I subscribe to the IAU circulars alerting me to the latest discoveries. The circulars get to me quicker and give more details, but the CBAT page is a backup in case I miss any.

The process is as follows:

1. Check CBAT's Recent Supernova page for already-discovered supernovae. Go back at least 12 months when checking. I check the whole list. That's why it's there. Check not only by galaxy name but also by its coordinates.

2. Check the Minor Planet Center page for known asteroids, at that location, at that time. This should be done for both the discovery image and the confirmation image.

3. Check past patrols. I NEVER throw images away. All of the patrols I have ever made are saved on CDROMs and archived. These are an invaluable resource when checking for possible variables, especially dwarf novae, in the field.

4. At this point I will stop one of the telescopes and go for at least one confirmation image. I try for multiple images, usually ten, if the weather allows. This process eliminates hot pixels and cosmic ray hits, which can cause starlike artifacts to appear.

5. I do astrometry on the suspect, on the discovery image and on the last confirmation image. If there is a difference of greater than one arc-second, I will wait and re-image as late as possible that night. This might be due to the suspect being an as yet uncharted asteroid.

6. I have recently added an additional check. I have created a list of "falsely reported" supernovae by others from the CBAT archive. This alerts me to fields where dwarf novae, variable stars and even HII regions have caused problems in the past. If the coordinates match even roughly, I become suspicious.

7. If all these things check out, I measure the offsets (the number of arc-seconds, e.g., north or south of the galactic nucleus) of where the suspect lies. I then prepare a preliminary report for CBAT.

8. A supernova has to be imaged and its position measured over at least two "periods of darkness" to ensure that it is not a moving object. This is usually over two nights but if the weather is likely to be a problem I will contact colleagues abroad, perhaps in the United States or Japan, and request that they take a confirmatory image. In these cases, the images might be less than a day apart but would still be acceptable if the astrometry showed no perceptible movement.

If all of these prove positive I will prepare a final report and draft release for the Central Bureau. Please see the following samples.

Preliminary Report

I wish to report a possible supernova in MCG+10-19-85. I have recorded this on multiple images on a single night. I will obtain a confirmation image on a second night as soon as the weather permits.
Tom Boles

SUPERNOVA IN MCG+10-19-85
T. Boles, Coddenham, England, reports the discovery of an apparent supernova (mag. 18.5) on an unfiltered CCD image taken on Feb 19.119 UT with a 0.35-m reflector. The new object is located at R.A. = 13h32m17s.38, Decl. = +60°23′43″.4, which is approximately 8″.3 east and 5″.8 north of the center of MCG+10-19-85. The suspect is not present on Boles' images from 2003 May 25 and Apr 02 (limiting mag. 19.5) and it is not present on DSS II red (1997.480) or blue plates (1995.310).

Final Report

I have succeeded in acquiring a confirmation image of the possible supernova in MCG+10-19-85. End figures for astrometry on the second night were 17s.39 and 43″.6. All the usual checks have been made. The MPC shows no asteroids in the area for either night.
Tom Boles

SUPERNOVA IN MCG+10-19-85
T. Boles, Coddenham, England, reports the discovery of an apparent supernova (mag. 18.5) on unfiltered CCD images taken on Feb 19.119 and 19.790 UT with a 0.35-m reflector. The new object is located at R.A. = 13h32m17s.38, Decl. = +60°23′43″.4, which is approximately 8″.3 east and 5″.8 north of the center of MCG+10-19-85. The suspect is not present on Boles' images from 2003 May 25 and Apr 02 (limiting mag. 19.5) and it is not present on DSS II red (1997.480) or blue plates (1995.310).

The above reports resulted in SN2004Z. See IAU Circular 8290.

Once this has been completed, the waiting begins. An IAU Circular might be released, giving the object's assigned designation and confirming its type (type I or II from spectroscopic analysis) or more likely, one circular will announce its possible discovery (all supernovae are "possible" supernovae until a spectrogram provides confirmation) and a second, a few days later, will give its type and characteristics.

Even if you have done your checks well, this can be a very stressful time. You need to realize that, based on what you have reported, a large professional telescope somewhere in the world might be taken off its current program and given

Figure 12.9. Coddenham Observatory looking east – a third Paramount resides inside the dome.

the task of acquiring a spectrogram of your suspect. It has to be an instrument of at least 1m or 1.5m in diameter to collect enough light for analysis. The Keck and the Hubble Space telescopes have been used on occasions for faint and interesting suspects. One could become very unpopular very quickly by submitting erroneous suspects. Professional astronomers have to fight long and hard for valuable telescope time. It is scarce enough without it being called on recklessly. If the tests are carried out religiously, this should never happen.

What Next?

Without doubt the biggest challenge for the future is the development of software to automatically check images. This is extremely difficult to do. Sky conditions vary, telescope pointing is never perfect and artifacts confuse even the cleverest checking algorithms. Some professionals have developed working programs to do this, but until it becomes available to amateurs we will need to continue with the slow, but reliable, process of manual checking.

Supernova patrolling is hard work. It can also be fun. I find it very satisfying. If I didn't, I would stop. It is a special feeling when you first see a suspect on your computer screen and you work through the checks. With each check your confidence grows until you are finally looking at an object that you are 99.9%

certain is a real supernova. The realization that you are probably the only person on the planet, and perhaps in the galaxy, to know of its existence provides the energy and motivation for the next night of supernova patrolling.

Related Web sites

Apogee Cameras	www.ccd.com/index.html
TheSky Suite	www.bisque.com/Products/TheSky/thesky.asp
CCDSoft	www.bisque.com/Products/CCDSoft/ccd.asp
Orchestrate	www.bisque.com/Products/Orchestrate/orchestrate.asp
TPoint	www.bisque.com/TPoint/tpoint.asp
Paramount	www.bisque.com/Products/Paramount/

British Astronomical Association
www.britastro.org
Royal Astronomical Society
www.ras.org.uk
Central Bureau for Astronomical Telegrams (CBAT)
http://cfa-www.harvard.edu/iau/cbat.html
List of Recently Discovered Supernovae
http://cfa-www.harvard.edu/iau/lists/RecentSupernovae.html

Contributors

David Ratledge is an IT Manager living in Lancashire, UK. He built his first telescope more than 40 years ago and soon became hooked on photographing the night sky. That long apprenticeship served him well and provided a sound footing to embark on digital imaging when it arrived on the amateur scene more than 10 years ago. Since then many of his images have been published on both sides of the Atlantic. He has already written or edited three books for Springer: *The Art and Science of CCD Astronomy*, *Software and Data for Practical Astronomers* and *Observing the Caldwell Objects*. He is a regular contributor to *Sky & Telescope* magazine, for which he is an associate editor. He is chairman of Bolton Astronomical Society.
Email: davidr@deep-sky.co.uk; Homepage: http://www.deep-sky.co.uk

David Haworth started astro-imaging with a Cookbook CCD camera he built in 1996, and since then he has used SBIG CCD cameras, digital cameras, digital SLR cameras, Web cameras, Meade LPI and 35mm film cameras to image the sky. David's images have appeared in *Lunar Photo of the Day* (LPOD), Astronomy.com Photo Gallery, Spaceweather.com, *Images of the Moon E-Book*, Orion Telescopes & Binoculars catalog and advertisements in both *Sky & Telescope* and *Astronomy* magazines. David is a member of the team of astronomers who produce the annual "Imaging the Sky" conference. He enjoys astronomical imaging and then processing his images to bring out details that cannot be seen by visual observing.
Email: David.A.Haworth@tek.com; Website: www.stargazing.net/david

Keith Wiley and Steve Chambers. Keith is a Ph.D. student studying computer science at the University of New Mexico. While he has followed the developments of webcam hardware modifications quite closely, he is primarily a computer programmer and, as such, spends much of his time writing image acquisition and image processing software oriented toward astrophotography-related imaging. **Steve** is a professional biochemist and keen amateur astronomer. His background in biochemistry has played an important role in the development of Steve's hobby by ensuring he never has quite enough money to buy off-the-shelf equipment. His DIY (do-it-yourself) astronomy projects include a half-meter computerized telescope and an automated sky survey. In 2001 Steve developed a method to allow modern webcams to be used for deep-sky astronomy – the rest, they say, is history!
Email: kwiley@cs.unm.edu & Steve@pmdo.com; Webpages: http://www.unm.edu/~keithw/ (Keith) http://www.pmdo.com/ (Steve)

Johannes Schedler has been a pioneer of deep-sky imaging with digital SLR cameras and several of his stunning color images, taken with a Canon D60, have been published in *Sky & Telescope* magazine. He works as technical manager (chemical engineer M.Sc.) for the Austrian company CTP Air Pollution Control. He lives south of Graz, the Styrian capital, and is a keen astro-imager using webcams, digital cameras and CCDs. He has been imaging now for more than 5 years. His Panther Observatory houses a Celestron C11 Schmidt-Cassegrain and a 4-inch APO refractor on a heavy permanent GEM.
Email: j.schedler@ctp.at; Homepage: *http://panther-observatory.com*

Christian Buil could justifiably be regarded as the father of amateur CCD astronomy. He was probably the first amateur to build a working CCD camera and publish details on its construction and operation, as long ago as 1987. Originally only available in French, this was later expanded and translated into English as *CCD Astronomy*, published by Willmann-Bell Inc. An essential part of CCD astronomy is image processing software, and here too Christian has been a pioneer with his programs: MIPS, Qmips and now IRIS. His main interest has always been spectroscopy; he is one of a French team who have access to the 0.6-meter Newtonian and 1-meter Cassegrain at the famous Pic du Midi Observatory to carry out their work. In his professional life he is an optical engineer specializing in opto-electronics with the French Space Agency in Toulouse.
Email: christian.buil@wanadoo.fr; Website: http://www.astrosurf.com/buil/

Damian A. Peach is now widely regarded as one of the world's foremost high-resolution planetary observers. His images have graced the covers of many books, magazines, professional publications and Web pages. He has appeared several times on BBC television, and for their Mars Spectacular he produced fabulous images of Mars live on air. He is an assistant director of the BAA Jupiter Section, the BAA Saturn Section and the ALPO Jupiter Section.
Email: dpeach_78@yahoo.co.uk; Website: http://www.damianpeach.com

Brian Lula, a mechanical engineer by profession, is fortunate that his hobby has become his career. He is president of the U.S. subsidiary of Physik Instrumente, a world-leading manufacturer of high-tech nano-positioning products for scientific research including astronomical applications. Brian provides a unique perspective to the discussion of astronomical CCD imaging as he has designed and built observatory-class amateur telescopes and equatorial mounts. Brian has been building telescopes for more than 30 years, including a 20-inch f/5 Newtonian astrograph optimized for wide-field CCD imaging, a CCD Cookbook 245 camera, several observatories and a host of other smaller telescopes and mounts. Brian's astronomical images have appeared many times in leading magazines and prestigious Web sites such as NASA's "Astronomy Picture of the Day." His images have also been shown on CBS's *Sunday Morning News* and were selected for an exhibition highlighting Brian's work on telescopes and imaging at the Smithsonian National Air and Space Museum in Washington, DC.
Email: brianl@heavensgloryobservatory.com; Website: www.heavensgloryobservatory.com

Robert Gendler has been an astrophotographer for almost a decade and has had images published in more than 200 magazines, books and calendars, including many front covers. He has achieved all this imaging from his home, in light-polluted suburbia, using creative imaging techniques. Twenty-six of his images have been featured on NASA's Web site "Astronomy Picture of the Day," which is more than any amateur astrophotographer in the world. He has also had his work featured on national TV such as the NBC *Today Show*, the CBS *Sunday Morning Show* and in an upcoming PBS *NOVA* documentary, "Cosmic Origins." His image of the Andromeda galaxy was selected as one of the greatest astronomical images of the last 30 years by *Astronomy* magazine in September 2003.
Email: robgendler@att.net; Homepage: http://www.robgendlerastropics.com

Dale E. Mais has been involved in amateur astronomy most of his life. He is an endocrinology researcher working for a bio-tech company in the San Diego area. While his biology and chemistry degrees serve him well in his professional life, it is his chemistry background that he enjoys applying to spectroscopy. He is fortunate to have an observatory with a Celestron 14 as its primary instrument, CCD cameras and an AstroPhysics 5.1-inch. His location, 12 miles from Mount Palomar, means he benefits from outstanding seeing and relatively dark skies, which the Hale telescope enjoys. His primary interest is spectroscopy and photometry of long-period variables of the Mira class. These are stars of type M, which evolve into S- and C-type stars and are of great interest because of the peculiar heavy-metal abundances, which get cooked up within them and appear at their surfaces. In particular, he is monitoring these type of stars for potential flare-up events, which can be followed up with moderate-resolution spectroscopy.
Email: dmais@ligand.com; URL: http://mais-ccd-spectroscopy.com

Tom Boles, who is originally from Glasgow, Scotland, spent many happy years as a telescope designer and maker for Charles Frank Ltd., from whom many amateurs in the UK acquired their first telescope. Today he is a retired computer and telecom support engineer, having held director-level positions with several multi-national equipment marketing and support companies. He is currently president of the British Astronomical Association, a Fellow of the Royal Astronomical Society, a member of the Webb Society and The Astronomer. His devouring passion is extragalactic supernovae: their physics, appearance, demographics and, of course, their discovery. He patrols on every available clear night and monitors some 12,000 galaxies for supernovae. In the last year his total supernova discoveries passed the 80 mark, and he is well on his way toward a target of 100. Since he started patrolling, he has recorded more than a third of a million galaxy images and clocked more than 10,000 hours collecting and searching images.
Email: tomboles@coddenhamobservatories.org; Website: www.coddenham-observatories.org